Winning the Next War

A volume in the series

CORNELL STUDIES IN SECURITY AFFAIRS

edited by Robert J. Art, Robert Jervis, *and* Stephen M. Walt

A full list of titles in the series appears at the end of the book.

Winning the Next War

INNOVATION AND THE MODERN MILITARY

STEPHEN PETER ROSEN

Cornell University Press

ITHACA AND LONDON

First published 1991 by Cornell University Press.
First printing, Cornell Paperbacks, 1994.

International Standard Book Number 0-8014-2556-5 (cloth)
International Standard Book Number 0-8014-8196-1 (paper)
Library of Congress Catalog Card Number 91-55235
Printed in the United States of America
*Librarians: Library of Congress cataloging information
appears on the last page of the book.*

⊗ The paper in this book meets the minimum requirements
of the American National Standard for Information Sciences—
Permanence of Paper for Printed Library Materials, ANSI Z39.48–1984.

Contents

Acknowledgments

Halfway through the work on this book, I got sick. Dr. Peter Mauch and Lynn Thompson, R.N., saved my life.

Al Bernstein built the Strategy Department at the Naval War College. While it lasted, it was the best in the business. He gave me an intellectual home in that department that made this book much easier to write.

I took my first university courses on national security affairs from Robert Jervis over fifteen years ago. He has read more draft chapters of this book and written more copious and useful comments about them than anyone else.

From James Q. Wilson I learned that an intellectual could gain a true understanding of bureaucratic politics by looking at them from the outside, although I did not fully realize this until I had left the university world and seen bureaucracies from the inside. From Andrew W. Marshall, I learned that it was possible to be a bureaucrat and an intellectual, in the finest sense of that word, and that it was important to understand bureaucratic politics. Were it not for Samuel P. Huntington, I would today be a bureaucrat, in the not so fine sense of the word, instead of an intellectual.

My wife, Mandana Sassanfar, has made me happier than I had any right to expect, happier even than writing this book did.

This book has flaws, but is as good as I could make it. I must acknowledge, therefore, that its shortcomings reflect shortcomings in my intellect and character.

Portions of the book are based on my article "Theories of Victory: Understanding Military Innovation," which appeared in *International Security* 13 (1988), and I am grateful for the journal's permission to rework this material.

S. P. R.

Winning the Next War

[1]

Thinking about
Military Innovation

When and why do military organizations make major innovations in the way they fight? Are they doomed always to "fight the last war"? Is it easier for them to innovate in peacetime, when the enemy is not engaging them in combat, or is innovation easier in wartime precisely because they can learn from combat? How does technological innovation in the military differ from innovation in the way a military organization fights? What is the role of intelligence in military innovation? These questions are central to the debate about how the United States can and should prepare for the military problems it faces. They are also the central questions of this book.

All social innovation is difficult. Machiavelli noted over four hundred years ago that "there is nothing more difficult to carry out, nor more doubtful of success, nor more dangerous to handle, than to initiate a new order of things. For the reformer has enemies in all those who profit by the old order, and only lukewarm defenders in all those who would profit by the new order . . . [because of] the incredulity of mankind, who do not truly believe in anything new until they have had actual experience of it."[1] Today, Machiavelli might be even more pessimistic, because the task of political innovation no longer involves social change alone. The particular problem facing men and women involved in the study and practice of modern politics is how to get bureaucracies to innovate.

[1] Niccolo Machiavelli, *The Prince* (New York: New American Library, 1952), book 6, "Of New Dominions Which Have Been Acquired by One's Own Arms and Ability," pp. 49–50.

Pessimism about bureaucratic innovation has led to despair and to the hope that it might be possible to ignore "the bureaucracy" and to substitute a small group of talented individuals for the bureaucracy when change is needed.[2] This might be possible in some areas of government. In recent times, men have tried to make foreign policy out of the White House, for example, in order to circumvent the State Department bureaucracy. But no one has yet explained how nations can wage war under modern conditions without operating with and through the huge bureaucracy that is the American military. The problem of military innovation is necessarily a problem of bureaucratic innovation.

Almost everything we know in theory about large bureaucracies suggests not only that they are hard to change, but that they are *designed not to change*. The first student of bureaucracies, Max Weber, argued that the essence of a bureaucracy was routine, repetitive, orderly action. Bureaucracies were not *supposed* to innovate, by their very nature.[3] Military bureaucracies, moreover, are especially resistant to change. Colonel John Mitchell of the British Army may have been one of the first men to try and explain why, when he wrote in 1839: "Officers enter the army at an age when they are more likely to take up existing opinions than to form their own. They grow up carrying into effect orders and regulations founded on those received opinions; they become, in some measure identified with existing views, till, in the course of years, the ideas thus gradually imbibed get too firmly rooted to be either shaken or eradicated by the force of argument or reflection. In no profession is the dread of innovation so great as in the army."[4]

Modern social scientists have elaborated on the inflexibility of military institutions. The horse cavalry persisted in European armies well into the twentieth century despite the impact of modern weapons because, one historian wrote, the "cavalry was a club, an exclusive one . . . a group of men who were at once hard-riding and hard

---

[2]See, for example, Henry Kissinger, "Domestic Structure and Foreign Policy," in Henry Kissinger, *American Foreign Policy* (New York: Norton, 1974), pp. 18–23; Lincoln P. Bloomfield, "Planning Foreign Policy: Can It Be Done?" *Political Science Quarterly* 93 (Fall 1978), 387–88.

[3]See the discussion of this point in Karl W. Deutsch, "On Theory and Research in Innovation," in Richard L. Merritt and Anna J. Merritt, eds., *Innovation in the Public Sector* (Beverly Hills: Sage, 1985), p. 20; Marshall W. Meyer, *Change in Public Bureaucracies* (London: Cambridge University Press, 1979), p. 99.

[4]John Mitchell, *Thoughts on Tactics and Military Organization,* cited in Jay Luvaas, *The Education of an Army: British Military Thought 1815–1940* (Chicago: University of Chicago Press, 1964), p. 43.

[2]

headed. . . . The cavalry was the home of tradition, the seat of romance, the haven of the well connected." Furthermore, it rarely had to go to war and so rarely had to test its romantic ideas against military realities. To some extent, this romanticism was necessary— no one would want a fighting force to go into battle that did not have confidence in its weapons. But this attitude did not foster innovation.[5]

Bureaucracies do innovate, however, even military ones, and the question thus becomes not whether but why and how they can change. Social scientists have spent much time and effort explaining bureaucratic innovation in general. The results have not been encouraging. As James Q. Wilson noted twenty-five years ago, "The process of organizational change is perhaps the least developed aspect of organization theory. . . . It has always seemed easier, and even more interesting, to analyze 'operational failures' to adapt."[6] Essays in the Herbert Simon's 1982 collection on organizational behavior similarly lamented that "little or nothing is known of the determinants of the rate and direction of inventive activity, of how technical know-how is 'stored' and reproduced by a society," and that "very few data are available to show whether such circumstances [such as hard times or vigorous competition] stimulate search activity and innovation."[7]

Much of the problem with social scientific studies of bureaucratic innovation has been that as one study found a factor that seemed to be associated with innovation, another would find evidence of innovation when that factor was absent, or even when the opposite of that factor was present. For example, Anthony Downs suggested that bureaucracies would innovate in order to acquire additional resources or to protect themselves from bureaucratic rivals or from changes in the environment that threatened to take resources away.[8] The implication was that when money and manpower were in short supply, bureaucracies would change their behavior to protect what they had or to get more. But a study of ninety-four public health agencies found a "surprising, but unequivocal" result, that "the re-

[5]Edward L. Katzenbach, Jr., "The Horse Cavalry in the Twentieth Century," reprinted in Robert J. Art and Kenneth Waltz, eds., *The Use of Force: International Politics and Foreign Policy* (Boston: Little, Brown, 1971), pp. 277, 281, 297.

[6]James Q. Wilson, "Innovation in Organization: Notes toward a Theory," in James D. Thompson, ed., *Approaches to Organizational Design* (Pittsburgh, University of Pittsburgh Press, 1966), p. 195.

[7]Herbert Simon, *Models of Bounded Rationality*, vol. 2, *Behavioral Economics and Business Organization* (Cambridge: MIT Press, 1982), pp. 394–95.

[8]Anthony Downs, *Inside Bureaucracy* (Boston: Little, Brown, 1967), pp. 198–200.

sources available . . . had on the average almost no impact on proportional innovation."[9] Richard Cyert and James March developed a theory of organizational behavior that predicted that failures in business firms would lead those firms to search for new organizational activities, but their survey of the real world behavior of twelve firms in four industries indicated that business failure did usually not lead to innovation, but only to more failure. They then speculated that success in business could lead to an accumulation of surplus resources within the firm, and that this "slack" would incline the firm to take more risks with innovation. This left the authors in the unfortunate position of predicting that "firms will innovate both when successful and unsuccessful."[10]

Some scholars looked at innovation from a different perspective. Instead of looking at the economics of innovation, they looked at innovation as a problem of organizational learning. These studies suggested that a loosely structured organization, unfettered by rigid internal hierarchies and overspecialization, would be more amenable to a free internal debate that would facilitate organizational learning and so facilitate change.[11] These studies did not suggest, however, that organizations that were good at free internal discussion would also be good at making up their minds and taking decisive action. Indeed, less hierarchy and specialization inclined organizations to internal deadlock and to ineffective action when a policy had to be implemented. Since innovation in a bureaucracy means actually doing something differently, not just having new ideas, this approach, of looking at innovation as a question of organizational learning, did not provide a complete explanation.

The inability of social science to explain innovation has not gone unnoticed by experts in the field of bureaucratic behavior. No good explanation of bureaucratic innovation exists. There are only contradictory results from different studies. One survey published in 1971 stated that academics had come up with thirty-eight different propositions about innovation, and that they disagreed about thirty-four of these. The four that were not the subject of controversy were

[9]Lawrence B. Mohr, "Determinants of Innovation in Organizations," *American Political Science Review* 63 (March 1969), 121.

[10]Richard Cyert and James March, *A Behavioral Theory of the Firm* (Englewood Cliffs, N.J.: Prentice-Hall, 1963), pp. 278–79.

[11]For examples of this analysis see Chris Argyris and Donald Schon, *Organizational Learning: A Theory of Action Perspective* (Reading, Mass.: Addison-Wesley, 1978), pp. 38–39, 143, 145; Lloyd A. Rowe and William B. Bose, "Organizational Innovation: Current Research and Evolving Concepts," *Public Administration Review* 34 (May/June 1974), 289.

the four which had not yet been discussed by academic experts.[12] Summing up the state of knowledge in the field of bureaucratic innovation, two scholars found that "factors found to be important for innovation in one study are found to be considerably less important, not important at all, or even inversely important in another study. This phenomenon occurs with relentless regularity."[13] The authors advised others to set aside a search for grand theories of innovation for the time being and to recognize that different kinds of innovation occur for different reasons in the same organization, and that different organizations will handle innovation very differently. This is the approach taken in the present volume.

What can be said about innovation in military bureaucracies? Following the advice of experts that it would be wise to break the phenomenon of innovation up into more manageable categories, this book divides it into problems of innovation in the operational behavior of the military during periods of peace, innovation in the operational behavior of the military in wartime, and technological innovation in the military. The first two focus on issues of human social behavior, the third on those of building new machines. I argue that though these problems are very different from each other, each is a reasonably coherent problem in itself. The advice of the experts, however, was not only to break the phenomenon of innovation down into smaller pieces, but to acknowledge that it is unlikely that explanations of innovation will have universal applicability. Therefore, because this book concentrates on successful cases of innovation within the American and British military, the explanations offered here may or may not explain innovation in very different cultures, such as that of the Soviet Union or Japan.

The book focuses on successful instances of innovation, not on failure to innovate, because in bureaucracies the absence of innovation is the rule, the natural state. What needs explaining is success, cases where, despite the obstacles, innovations did occur and were put into practice. As I suggested above, there is already a well-developed set of studies explaining why military bureaucracies do not innovate. Understanding successful American innovations, however, requires some comparison with cases in which innovations were attempted but not put into effect. For this reason, examples of

[12]E. Rogers and F. Schoemaker, *Communication of Innovations* (New York: Free Press, 1971), cited in John Agnew, ed., *Innovation Research and Public Policy* (Syracuse: Syracuse University Press, 1980), pp. 75–100.
[13]George W. Downs, Jr., and Lawrence B. Mohr, "Conceptual Issues in the Study of Innovation," *Administrative Science Quarterly* 21 (December 1976), 700.

Table 1. Twenty-one innovations examined in this book

	United States	Great Britain
PEACETIME	Amphibious warfare, 1905–1940 Carrier aviation, 1918–1943 Helicopter airmobility, 1944–1965 Counter-insurgency, 1960–1967	Carrier aviation, 1918–1940 Air defense, 1916–1940

WARTIME	Jungle warfare, 1942–1943
	
	Strategic bomber targeting,
	    1941–1944
	
	Submarine warfare, 1941–1945
	Long-range escort fighter,
	    1940–1944
		Jungle warfare, 1939–1944
		Amphibious warfare, 1914–1915
		Strategic bomber targeting,
		    1939–1945
		The tank, 1914–1918

TECHNOLOGICAL	Guided missiles, 1918–1956
	Proximity fuse, 1941–1944
	Electronics warfare, 1921–1945
	Centimeter wave radar,
	    1930–1942
	Ordnance, 1918–1945
		Electronics warfare, 1938–1945
		Centimeter wave radar,
		    1938–1942

failed innovations, primarily drawn from the history of the British military, are used when they provide "natural experiments" that test or help refine the explanations for innovation I advance. These are cases in which the British tried to innovate in the same areas as the United States, faced many of the same problems, but failed because their behavior differed in a few key areas. Those areas of difference may help to isolate the causes of successful innovation.

Twenty-one cases of military innovation (listed in Table 1) both successful and unsuccessful, are studied in this book. Ten of these are examined in some detail. Other recent studies of military innovation have been limited to two or three cases.[14] These examples

[14]Barry R. Posen, in *The Sources of Military Doctrine: France, Britain, and Germany between the World Wars* (Ithaca: Cornell University Press, 1984), studies successful innovation in the Royal Air Force between the world wars, and noninnovation in the French and German armies in the same period. Matthew Evangelista, in his book *Innovation and the Arms Race: How the United States and the Soviet Union Develop New Military Technologies* (Ithaca: Cornell University Press, 1988), focuses on the development of tactical nuclear weapons in the two countries.

were selected for one of two reasons. Many represent innovations that were simply the most important in the history of the particular military organization. The development of carrier aviation in the U.S. Navy, amphibious warfare in the U.S. Marine Corps, and strategic bombing and the ballistic missile in the U.S. Air Force fall obviously into this category. Other cases are less important but help us test existing theories of innovation. The development of air defenses in Great Britain and helicopter aviation in the U.S. Army, for example, are often cited as proof of the theory that military innovation is prompted by civilian intervention in the process of military development. The development of strategic bombing capabilities in the U.S. Air Force in World War II is often cited as proof that military organizations innovate only in wartime after catastrophic failure.

One obvious case of innovation is not studied in this book: the development of the first nuclear weapons. For reasons that I will explain, this innovation was unique, and is outside the scope of the explanations laid out in this book.

The focus on the British and American militaries precludes a genuine comparative study of military innovation in all modern military bureaucracies. Such a study would be enormously valuable. In practice, however, an analysis of military innovation requires a detailed look at the internal operations of a particular organization, which requires access to declassified records, internal administrative histories, intelligence reports, and so on. The necessary research work for an international comparative study would be enormous.

Following this chapter, the book is divided into three major sections corresponding to the three categories of innovation: peacetime, wartime, and technological. In the first two, a major innovation is defined as a change in one of the primary combat arms of a service in the way it fights or alternatively, as the creation of a new combat arm. A combat arm is a functional division within the military in which one weapon system dominates the way in which its units fight. Within the U.S. Army, for example, one can identify the infantry, artillery, armor, and helicopter aviation combat arms. A major innovation involves a change in the concepts of operation of that combat arm, that is, the ideas governing the ways it uses its forces to win a campaign, as opposed to a tactical innovation, which is a change in the way individual weapons are applied to the target and environment in battle. A major innovation also involves a change in the relation of that combat arm to other combat arms and a downgrading or abandoning of older concepts of operation and possibly

[7]

of a formerly dominant weapon. Changes in the formal doctrine of a military organization that leave the essential workings of that organization unaltered do not count as an innovation by this definition.[15]

However, a military innovation may not involve behavioral changes in the organization at all, but rather the creation of new military technologies. While this is quite distinct from the kind of innovation outlined above, a book on military innovation that ignored, for example, the creation of guided missiles, radar and electronics warfare, and so on, would hardly be complete. Thus, peacetime, wartime, and technological innovation in the military involve three distinct sets of intellectual and practical problems.

THE PROBLEM OF PEACETIME MILITARY INNOVATION

Almost every government bureaucracy has a function it executes on a day-to-day basis. Military organizations, in contrast, exist in order to fight a foreign enemy, and do not execute this function every day. Most of the time, the countries they serve are at peace. Military organizations plan and prepare for war, but they do not fight. Instead of being routinely "in business" and learning from ongoing experience, they must anticipate wars that may or may not occur. In addition, they are governed by professional officer corps into which new blood can only be introduced from below, and only with the approval of the senior leadership. These are the two aspects that define the problem of innovation in the peacetime military. What explanations have been suggested for peacetime military innovation?

The first that comes to mind is, of course, the conventional wisdom that catastrophic military defeat provides the catalyst that leads to change after the war ends. A moment's reflection on the two

[15]Kevin Sheehan, for example, reviewed three major changes in U.S. Army doctrine for ground combat from 1945 to 1980 and found that although formal doctrine shifted—from the Pentomic Division concept to Active Defense to Airland Battle—the army's continuing focus was on fighting a conventional war with the Soviet army on the battlefield of central Europe. Despite formal acknowledgment of the need to innovate in response to nuclear battlefield capabilities and to develop combat abilities for theaters outside Europe, the central combat function of the army remained unchanged. See Kevin Patrick Sheehan, "Preparing for an Imaginary War: Examining Peacetime Functions and Changes of Army Doctrine" (Ph.D. diss., Harvard University, 1988), pp. 352–56. The special case of innovation in U.S. Army helicopter aviation will be examined in chapter 3 below.

[8]

problems set forth above, however, is enough to suggest that defeat in war is neither necessary nor sufficient to produce innovation. Defeat by itself does not tell a military organization what future wars will look like, only that its preparations for the war just ended were not adequate. Defeat by itself does not insure that significantly new talent will rise to the top of the officer corps, since the postwar officer corps may well be composed of prewar junior officers trained in prewar methods. History is in fact full of examples of armies and navies that were defeated and went on being defeated because they did not innovate. The czarist army after the Russo-Japanese War was not particularly innovative,[16] and the U.S. Army did not rush to develop innovate capabilities for counterinsurgency after the Vietnam War.[17] At the same time, there have been a number of notable examples of military organizations that have managed to innovate without having suffered defeat. Chapter 2 will lay out in some detail how the U.S. Army, Navy, and Marine Corps developed new capabilities for helicopter warfare, carrier aviation, and amphibious assault after World Wars I and II, wars in which not only was the United States victorious but also each of the services in question fulfilled its mission successfully. For now, it is enough to note that failure in war has not been necessary or sufficient for peacetime innovation.

A better explanation has been offered, one that follows from the observation that because of the peculiar character of military organizations in peacetime they are simply unlikely to innovate at all if left to themselves: military innovation must be the result of civilian intervention. Citing military hierarchy, traditionalism, romanticism, and the lack of peacetime tests of effectiveness, Kurt Lang has argued that military innovations "are largely promoted by civilians, who have often shown themselves more sensitive to changing needs than the professional military leadership. Military innovators like General [William] Mitchell . . . and Rickover . . . have frequently been blocked by their more tradition-oriented colleagues."[18] More

---

[16]See, for example, John Bushnell, "The Tsarist Army after the Russo-Japanese War: The View from the Field," in Charles R. Shrader, ed., *Proceedings of the 1982 International Military History Symposium: The Impact of Unsuccessful Military Campaigns on Military Institutions 1860–1980* (Carlisle Barricks, Pa.: U.S. Army War College, 1982), pp. 77–99.

[17]John P. Lovell, "Vietnam and the U.S. Army: Learning to Cope with Failure," in George Osborn, ed., *Democracy, Strategy, and Vietnam* (Lexington, Mass.: Lexington Books, 1987), pp. 121–54.

[18]Kurt Lang, "Military Organizations," in James G. March, ed., *Handbook of Organizations* (Chicago: Rand McNally, 1965), p. 857.

recently, Barry Posen has concluded after studying the British Royal Air Force, the French army, and the German army, all in the period between the two world wars, that "we see little internally generated innovation in the three cases." The French army failed to innovate, Posen argues; the changes in the German army did not represent genuine innovation; and the Royal Air Force innovated only after civilian leaders had administered an external shock.[19] In a variation of Lang's argument, Posen suggests that civilian intervention produces military innovation in peacetime, either directly or indirectly, through officers he calls military "mavericks," who provide civilians with the military expertise they lack.

Civilian intervention is an appealing *deus ex machina* that might explain innovation in peacetime military bureaucracies. But observations of the difficulties civilian leaders, up to and including the president of the United States, have had in bending the military to their desires should again lead us to be cautious. Richard Neustadt has listed five conditions that must prevail if a president's order is to be readily obeyed by his bureaucratic subordinates. The president himself must be clearly involved in the decision, and his order must be unambiguous. His order must be widely publicized, and "the men who receive it [must have] control over everything needed to carry it out," and they must have no doubt of his "authority to issue" the order.[20] How well are these conditions satisfied when the president issues an order to the military to innovate? His order may not be unambiguous. President Lyndon Johnson knew by 1967 that he was not content with the way the American military was fighting the war in Vietnam. He wanted military innovations to help win the war. But he could not give unambiguous orders because he did not know exactly what he wanted: "The President . . . asked General [Harold] Johnson [chief of staff, U.S. Army] to have the Joint Chiefs 'search for imaginative ideas to bring this war to a conclusion.' He said that he did not want them to just recommend more men or that we drop the Atom bomb. The President said he could think of those ideas. The President asked Johnson to have the Joint Chiefs come up with some new programs."[21]

The order to innovate is likely to be ambiguous because what is being ordered is not some familiar, well-defined task, but something

[19]Posen, *Sources of Military Doctrine*, pp. 224, 226.

[20]Richard Neustadt, *Presidential Power* (New York: John Wiley, 1980), p. 16.

[21]Minutes of 12 September 1967 weekly luncheon with Secretaries Dean Rusk and Robert McNamara, Walter Rostow, George Christian, Harold Johnson, Jim Jones, notetaker, Declassified Documents Registry Service, 1987, #1798.

that has never been done before. Those being ordered to innovate may well not have control over everything needed to carry out the order, particularly if what is needed is unconventional creativity. And the professional military may well regard an order to fight differently as being outside the legitimate authority of civilian leaders. Military professionals, not civilian politicians, are supposed to be the repository of expert knowledge on how to fight. The classic example of such a reaction can be found in the failure of the U.S. Army to develop army-wide capabilities for counterinsurgency even after being personally ordered to do so by the president. John F. Kennedy held a meeting with senior military officers to make clear his personal interest in counterinsurgency, and followed up on the meeting by creating a committee on which his brother Robert sat. Nonetheless, the army, because it believed in the superiority of conventionally trained infantry and that conventional wars would continue to dominate the army's strategic requirements, effectively blocked the shift to counterinsurgency while giving lip service to the president's orders.[22] In short, a civilian command to carry out military innovation is by its nature extremely difficult to enforce.

How do matters change if an unconventional military go-between, a "military maverick," intervenes between civilian leaders and the military bureaucracy? Posen does not clearly define the term "military maverick." If by "maverick" he simply means a military man who favors innovation more than the average officer, innovation will, by definition, by supported by mavericks. If, however, he is using "maverick" in its dictionary sense, he is referring to isolated and masterless men who have rejected the authority of their nominal military superiors. Such men have military credentials but are unwilling to acknowledge traditional authorities in the military. They "buck the chain of command" and appear over the heads of their military superiors in order to use an outside force to bring about the innovations they desire. This type of military man, working with civilian politicians, is the alleged engine of military innovation.

This explanation has great intuitive appeal for many people acquainted with military organizations, including at least one senior American officer.[23] The word "maverick" conjured up images of

[22]See Andrew Krepinevich, *The Army and Vietnam* (Baltimore: Johns Hopkins University Press, 1986), chap. 2.

[23]See General Donn A. Starry, U.S. Army, "To Change an Army," *Military Review* 63 (March 1963), 20–27, in which he argues that mavericks in the German Army between the world wars worked out a combined arms concept of fast-paced tank warfare.

Billy Mitchell and Hyman Rickover in the American military and Charles de Gaulle and B. H. Liddell Hart in the French and British militaries pushing tradition-bound organizations to innovate with the aid of highly placed civilian allies. But in examining their actions, historians have generally concluded that these extraordinary men were either less successful as innovators or less crucial in producing innovations that their public images suggest.

"Hap" Arnold was a fiery advocate of air power in the 1920s, and went on to command the U.S. Army Air Forces in World War II. He was an acknowledged disciple of Billy Mitchell in his fight for air-power in the 1920s and supported Mitchell throughout the struggles that culminated in his court martial. Mitchell was a maverick by anyone's definition, appealing to Congress and to public opinion for an independent Air Force in defiance of military discipline. Yet, when reflecting on Mitchell's career, Arnold commented: "In retrospect, I do not believe the War Department, as an agency, profited much, if at all, from the Mitchell 'period of influence on air development.' They seemed to set their mouths tighter, draw more into their shells, and, if anything, take a more narrow point of view of aviation as an offensive power in warfare."[24]

Hyman Rickover was an extraordinary military manager, even before he became involved with nuclear propulsion, and was willing to push himself and his associates into detailed study and perfectionist development of military technology. Yet all the detailed studies of his career agree that the image of "Rickover against the navy," with Rickover, supported by senators and congressmen, forcing the navy to accept nuclear propulsion over its own objections, was a myth deliberately created by Rickover. Nuclear propulsion was obviously in the interest of the navy, particularly the submarine force, and senior navy officers supported it before Rickover emerged as its most visible advocate, and they supported Rickover despite, and not because of, his aggressive self-promotion and cultivation of an independent role.[25]

[24]H. Arnold, *Global Mission* (New York: Hutchison, 1951), p. 97.

[25]An example of the Rickover myth in action can be found in the statements of Senator Henry Jackson cited in Michael Armacost, *The Politics of Weapons Innovation: The Thor-Jupiter Controversy* (New York: Columbia University Press, 1969), pp. 65–66. The best accounts of the development of nuclear propulsion in the navy and Rickover's role as one of several important navy figures can be found in Richard G. Hewlett and Francis Duncan, *Nuclear Navy 1946–1962* (Chicago: University of Chicago Press, 1974), pp. 27, 41, 56–58, 68–72, 79–85; Norman Polmar and Thomas Allen, *Rickover: Controversy and Genius* (New York: Simon and Schuster, 1982), pp. 118–34. A

Outside the United States, both Liddell Hart and de Gaulle were advocates of innovation in ground warfare, particularly in the field of tank warfare. Both took their cases for innovation outside military channels to the civilian chiefs of their war departments, to Leslie Hore-Belisha in the case of Liddell Hart and Paul Reynaud in the case of de Gaulle. As with Mitchell, the judgment of history is that by doing so they probably reduced the willingness of the professional military to consider innovation. In the words of one study, "in practice, 'outsiders' can seldom exert a direct influence on military reform because they lack full knowledge of the difficulties and options available. . . . On the other hand, the responsible military authorities tend to be all too well aware of the problem and [willing] to accept that only piecemeal or compromise measures are feasible."[26]

The contention that in peacetime military innovation is generated by civilians acting together with military mavericks is supported most persuasively by the development of anti-aircraft defenses in the Royal Air Force in the two years before the outbreak of World War II. Posen, along with several official British histories, points to the heavy emphasis placed on the development of the bomber and of offensive concepts of operation in the RAF after World War I, and to the relative neglect of the use of fighters for the air defense of Great Britain. He argues that it was not until civilians intervened in the military late in 1937 that the balance of resources devoted to fighters was redressed: "Prompted by systemic pressures, constraints, and incentives, civilians refashioned the doctrine of the Royal Air Force, elevating air defense to a temporary position of primacy. The persistent beliefs in the dominance of the aerial offensive became a factor to be overcome. The RAF's claims for the utility of air defense were ignored. The civilians knew what they wanted and set about getting it."[27]

Posen gives primary credit for this innovation to the civilian minister for the coordination of defence, Sir Thomas Inskip, but also

---

comparison of Rickover, J. Edgar Hoover, and Robert Moses can be found in Eugene Lewis, *Public Entrepreneurship: Toward a Theory of Bureaucratic Political Power* (Bloomington: Indiana University Press, 1980).

[26]Brian Bond and Martin Alexander, "Liddell Hart and De Gaulle: The Doctrine of Limited Liability and Mobile Defense," in Peter Paret, ed., *Makers of Modern Strategy* (Princeton: Princeton University Press, 1986), p. 623. For a balanced view of the impact of Liddell Hart's press campaign on behalf of a modern "mechanical force" in the British Army, see Harold R. Winton, *To Change an Army: General Sir John Burnett-Stuart and British Armored Doctrine, 1927–1938* (Lawrence: University Press of Kansas, 1988), p. 78.

[27]Posen, *Sources of Military Doctrine*, p. 175.

acknowledges the role of the RAF air chief marshall, Sir Hugh Dowding, the head of Fighter Command. Dowding is described as the sort of "maverick" who can be utilized by civilians to provide them with the military expertise they lack and who was protected by them from the RAF, which otherwise would have suppressed both Dowding and the development of air defenses.

There is no question that the leadership of the RAF placed more emphasis on the development of bombers than on air defense in the 1920s and 1930s. There is also little question that civilian intervention in the late 1930s eventually altered the balance of resources in favor of fighter aircraft, although there is some doubt whether these civilians had delineated for themselves exactly the kind of military doctrine they desired. Civilian officials wanted a large air force to deter Hitler, without spending a lot of money, and fighters were cheaper than bombers.[28] Moreover, there is great deal of doubt that Dowding was a "maverick" surviving in the RAF only because he had civilian protection. Though not well liked, Dowding had been chosen in 1930 by his RAF superiors, not civilians, to be the air member for supply and research on the Air Council, from which position he had successfully championed modern fighter aircraft development.

Nor was Dowding an isolated defender of fighters and air defense within the RAF. In the early 1930s, the two chiefs of the Air Staff who succeeded Sir Hugh Trenchard, the founding father of the RAF, were RAF officers who were also advocates of improved defensive capabilities. One, Sir Geoffrey Salmon, had served as air officer commander in chief of Air Defence Great Britain.[29] Under Trenchard's successors decisions were made to abandon biplanes in favor of high speed monoplanes with enclosed cockpits that incorporated oxygen supplies for pilots, retractable landing gear, and eight machine guns mounted in the wings, enough to afford a chance of shooting down a high speed target. These technological capabilities

---

[28]On the economic rationale for civilian preference for fighters over bombers, see Sir Charles Webster and Noble Frankland, *The Strategic Air Offensive against Germany 1939–1945* 4 vols. (London: Her Majesty's Stationery Office, hereinafter HMSO, 1961), hereinafter *SAOG*, 1:77; and G. C. Peden, *British Rearmament and the Treasury 1932–1939* (Edinburgh: Scottish Academic Press, 1979), p. 134. For the shift in aircraft production, see Basil Collier, *The Defence of the United Kingdom* (London: HMSO, 1957), p. 68; Richard Overy, *The Air War 1939–1945* (New York: Europa Press, 1980), pp. 32–33, 120; Derek Wood and Derek Dempster, *The Narrow Margin* (New York: McGraw-Hill, 1961), p. 462; Williamson Murray, *The Change in the European Balance of Power* (Princeton: Princeton University Press, 1984), p. 273.

[29]Francis K. Mason, *Battle over Britain* (London: McWhirter Twins, 1969), p. 80.

were specified by an RAF captain in July 1934, and incorporated directly into the Hurricane and Spitfire fighter aircraft.[30] Decisions were also made in this period to create "shadow" aircraft factories that could be rapidly mobilized and converted from automobile to aircraft parts production and to set up a volunteer reserve, so that additional pilots would be available to man the additional fighter planes that would be produced after mobilization.[31]

But the success of Britain's peacetime effort to create an air defense network was more than a matter of manpower and aircraft. It also depended on the rapid integration of radar as part of the system. They key to taking a technology that was first demonstrated in 1935 and creating in a short time a fully operational, battle-ready system appears to have been not civilian intervention but a long-standing interest on the part of the RAF leadership in the problems of command, control, communications, and intelligence in support of defensive fighter operations. The major studies of the creation of air defense in Great Britain have looked primarily at the role that civilians played in the invention of radar, and from that have concluded that civilians played the critical role, in a military innovation. However, they erred in not closely examining the activities of the Royal Air Force in the area of air defense before 1935.[32] From the early 1920s on, the RAF had been interested in developing the doctrinal and technological infrastructure needed to acquire, process, and distribute information about enemy aircraft to friendly forces. Thus the development of radar simply added a very superior form of battle intelligence to a system already set up to process and distribute information. It was a round technological peg going into a round doctrinal hole.

Reports prepared for the Committee on Imperial Defence from 1923 on do show an emphasis on offensive capability. Enemy bomber bases would themselves be attacked by British bombers, so that they could not be used to launch attacks on the Home Islands.

---

[30]Wood and Dempster, *Narrow Margin*, pp. 87–88.

[31]Ibid., p. 89, 96; Mason, *Battle over Britain*, pp. 97–98.

[32]For example, the classic study of one of the leaders of Britain's military-science establishment, Ronald W. Clark's biography of Sir Henry Tizard, makes extensive use of Tizard's and Lord Cherwell's papers, but not of any military documents, and concludes, not surprisingly, that civilians were responsible for the invention of air defense. See Clark, *Tizard* (Cambridge: MIT Press, 1965), pp. 431–36. More recently, Malcolm Smith's otherwise thorough book has essentially no data on the air defense activities of the Royal Air Force before 1938 and claims that the RAF had done little to prepare for this mission before British civilians became involved in RAF planning. See Malcolm Smith, *British Air Strategy between the Wars* (Oxford: Clarendon Press, 1984), p. 317.

Fighter strength would be kept to the minimum, to leave resources for the creation of the strongest possible bomber force. First the French and then from 1935 on, the Germans were seen as the primary air threat.[33] But defensive capability was endorsed as well, and there was thoughtful discussion of its essential elements. The commander of the British air defense command in World War I, General E. B. Ashmore, identified the key problem for air defense. He saw that air defense could not be solved simply by improving defensive aircraft.[34] What was needed was the construction of a system of ground control and communication that could direct fighters toward incoming bombers. This idea was incorporated into Britain's defense plans. The same 1923 RAF report that emphasized bomber capability also pointed to the difficulty that fighter aircraft would have in intercepting incoming bombers. The solution, it was argued, was to station fighters in the proper locations and more important, to get information on the location of the bombers to the fighters: "A highly organized system is essential for the rapid collection and distribution of information and intelligence regarding the movements of friendly and hostile aircraft throughout the whole area of possible air operations. In the opinion of the Sub-committee this organization is of such vital importance that they recommend its examination in detail, separately, at a later date."[35]

The chief of the Air Staff, Hugh Trenchard, singled out this recommendation for endorsement and announced that a joint subcommittee on the subject had been created. Subsequent reports endorsed by Trenchard indicate that a system supporting both fighter aircraft and antiaircraft artillery had been set up by 1928, utilizing ground observers and dedicated telephone lines.[36] Just as important was the system for managing information quickly during battle, so

[33]See Home Defence Sub-committee, Report 118A, "Continental Air Menace: Anti-Aircraft Defence," May 1923, and Report 205A, "Re-Orientation of the Air Defence System of Great Britain," April 1935, report by the reorientation subcommittee, Committee on Imperial Defence microfilm (hereinafter CID), CAB 3, Harvard University microfilm collection.

[34]"The great principle of air defense was not yet [in 1916] sufficiently recognized: that although aeroplanes are the first means of defense, they are ineffective unless supported by a control system on the ground." E. B. Ashmore, *Air Defence* (London: Longmans, Green, 1929), p. 39.

[35]Report 118A, "Continental Air Menace," p. 4.

[36]Home Defence Sub-committee, Report 135A, Subcommittee on Air Raid Precautions," June 1925, p. 7, and Report 160A, "Inter-Relationship between Authorities Concerned with the Defence of Great Britain Against Air Attack," March 1928, p. 2; CID, CAB 3, Harvard University microfilm collection.

as not to overload officers directing local air defense with reports not directly relevant to their tasks. The solution was proposed in 1929 by a committee chaired by Commander L. E. H. Maund of the Royal Navy. A filter system would be set up that would transmit only the information relevant to the fighting in one sector to the air officer commanding that sector. Complete information necessary for the strategic management of the air forces of Great Britain would be sent to the overall defense command in London. Maund concluded that "the application of the foregoing recommendations would result in the establishment in Air Defence Headquarters in London of a plotting table showing a comprehensive survey of all enemy air activities over this country." To reduce delays in utilizing information, a standardized system for reporting the number, course, and heading of enemy aircraft was instituted, and a complementary plan for telecommunications was developed by a separate Air Defence Land Line Telephone Sub-committee.[37] Subsequent reports in the early 1930s reemphasized the importance of rapid communication, and noted that improvements in radios carried by aircraft made more flexible deployments possible.[38]

Given this concern with the collection, management, and distribution of air defense intelligence, it is not surprising that when a new sensor was developed that promised better intelligence, it was seized on and developed with great speed. Starting in April 1935, shortly after the first British experiments with radar for aircraft detection, the Committee on Imperial Defence discussed the need for increased resources for the work of the Anti-aircraft Research Subcommittee and for the air intelligence system to take advantage of the "aircraft location and detection experiments."[39] RAF training facilities for radar operators were established at the end of 1936, and the construction of radar stations was accelerated in 1937.[40]

Not everything went smoothly. Funds for Air Defence Operations Center, in particular for a hardened operations room for Fighter Command, were slow in materializing, and improvements in communication lines made necessary by the use of radar were not im-

---

[37]Home Defence Sub-committee Report 167A "Coordination of Air Defence Intelligence," April 1929, pp. 5, 6, 10, CID, CAB 3, Harvard University microfilm collection.

[38]Report 205A, "Re-Orientation," p. 8.

[39]Ibid., p. 20; Report 209A [a sequel to 205A], "Re-Orientation of the Air Defence System of Great Britain," 23 September 1935, pp. 15–17, CID, CAB 3, Harvard University microfilm collection.

[40]Collier, *Defence of Great Britain*, p. 39.

mediately carried out.[41] But radar was incorporated successfully and rapidly into British air defenses. This was not because of civilian intervention, but because the RAF had laid a sound intellectual and organizational foundation for the use of this new technology. As one historian has recently noted: "The planning of air defence was not a major aspect of policy with with RAF's first decade. . . . However, in carefully establishing the essential constituents of an air defence system, and recognizing that its success depended, crucially, on the inter-relationship and cooperation of these elements, the [RAF] planners provided a sound framework for the future."[42] Thus steady doctrinal development within the military, not intervention by civilians or military "mavericks," explains this peacetime innovation.

Neither defeat nor civilian intervention adequately explains why or how military organizations innovate in peacetime. In looking for an explanation, we are forced to go back to the beginning. How should we think about military organizations in peacetime? Academics who have argued that military organizations tend to stagnate except when goaded by some outside force, have tended to make some implicit assumptions about the nature of military organizations, typically portraying them as monolithic units pursuing or protecting their self-interests, which are defined in fairly narrow terms—bureaucratic "turf," autonomy, protection of positions of power within a hierarchy, or marshalling of material resources. So it is a monolithic unit, "the French army," that failed to innovate in the interwar years, or another monolithic structure, "the Royal Air Force," that resisted air defense. To the extent that smaller units within these services are recognized by academics, they are either seen as undifferentiated groups that also pursue narrowly defined bureaucratic interests or as outsiders, "mavericks" that are not really part of the military at all and that owe their allegiance to the civilian world. In the classic model of bureaucratic politics advanced by Graham Allison in *Essence of Decision*, these subunits struggle or bargain with each other to obtain more of what they already have.[43] There is no necessary connection between their struggles and more or less innovative military policy, although the natural tendency might be

[41]"Note by Minister for the Coordination of Defence," 10 February 1939, p. 1, CID, CAB 3, Harvard University microfilm collection; R. V. Jones, *The Wizard War: British Scientific Intelligence 1939–1945* (New York: Coward, McCann, and Geohegan, 1978), p. 35.

[42]Neil Young, "British Home Air Defence Planning in the 1920s," *Journal of Strategic Studies* 11 (December 1988), 507.

[43]Graham Allison, *Essence of Decision: Explaining the Cuban Missile Crisis* (Boston: Little, Brown, 1971), pp. 162–81.

[18]

for this bargaining process to minimize any change that would injure any of the parties.

The search for explanations of military innovation during peacetime could begin with a different view of military organizations. Looking at the armed forces of the United States, for example, one notices that each service is far from monolithic and is not composed of subunits simply pursuing their own organizational self-interests. U.S. Army officers may come from the infantry, artillery, armor, aviation, airborne, or special forces. Navy officers may be carrier pilots from the fighter or attack communities, antisubmarine warfare pilots, submariners, surface ship commanders, or from an amphibious force. Each branch has its own culture and distinct way of thinking about the way war should be conducted, not only by its own branch, but by the other branches and services with which it would have to interact in combat. If the military organization is healthy, there is some general agreement among the various branches about how they should work together in wartime. This agreement is a dynamic condition. There is no permanent norm defining the dominant professional activity of the organization. Many theories concerning the relative priority of roles and missions compete. There are not only fights about the relative resources of each branch but also arguments about what the next war will or should look like. At times, these arguments can challenge the basic agreements as to how the services should operate in wartime.

If we start with this perspective, we will be inclined to regard military organizations as complex political communities in which the central concerns are those of any political community: who should rule, and how the "citizens" should live. The fundamental political values involved in these questions are from time to time subject to debate within the community. Military organizations have this political character to a greater degree than other bureaucratic organizations because military organizations are more divorced from the rest of society. An officer becomes an officer at an early age, in many instances after education at a service academy, and rises through promotion within the officer corps; an officer cannot transfer in as a major or colonel from civilian life, except in special circumstances. Military organizations govern almost every aspect of the lives of the members of their community. In wartime they determine who will live and die, and how, who will be honored, and who will sit on the sidelines.

If military innovation during peacetime is examined from this point of view, several ideas emerge. Because the service is a political community, innovation does not simply involve the transfer of resources from one group to another. It requires an "ideological" struggle that redefines the values that legitimate the activities of the citizens. Because the service is a military organization, and because it is victory in war that ultimately legitimates any military organization, this ideological struggle will revolve around a new theory of victory, an explanation of what the next war will look like and how officers must fight if it is to be won.

This new theory must be translated into concrete, new tasks that are performed every day, in peace and in war. New critical missions must be defined, so that soldiers can understand that good or bad performance of these tasks will be the measure of an officer's effectiveness and will determine, by reward and punishment, the way the community lives. If the new theory of victory is the new ideology, then the new tasks and performance measures are the legislation that transform the ideology into government. Without the development of new critical tasks, "ideological" innovations remain abstract and may not affect the way the organization actually behaves.

Ideology and values affect the distribution of power within a community, but they are only part of the process by which new leaders acquire power. Every form of government has its organizing principle that determines how power is acquired.[44] In military organizations, political power is not based on fear or virtue and is not won or lost by holding an election or by staging a coup d'etat. Military organizations are disciplined, hierarchical bureaucracies. Power is won through influence over who is promoted to positions of senior command. Control over the promotion of officers is the source of power in the military. The intellectual struggle concerning theories of victory is not irrelevant to that control, but must be supplemented by a hard-headed, concerted effort to gain control of whatever mechanisms determine who becomes an admiral or general. In peacetime, senior officers in the American military cannot be thrown out en masse. They are replaced individually, as men are systematically promoted from below. The organizational struggle that leads to innovation may thus require the creation of a new promotion pathway to the senior ranks, so that young officers learning and practicing the new way of war can rise to the top, as part of a generational change. The new pathway may be necessary to ensure that

[44]See, for example, Montesquieu's discussion "Of the Principles of the Three Kinds of Government," in Montesquieu, *The Spirit of the Laws*, trans. Thomas Nugent (New York: Hafner, 1949), vol. 1, book 3, pp. 19–28.

new skills are not relegated to professional oblivion. If a new skill is defined as a technical specialty, the officer with that skill will not be seen as having the broad background that qualifies him for the rank of general or an admiral.

These new promotion paths can only be created by senior officers who have political power within the service. Unless ranks are abolished or radical forms of civilian control are introduced, in a disciplined professional army senior officers will control the professional lives of junior officers. Change will come about through the actions of those who have the power. No "maverick," by definition, has that kind of power. A maverick is an outsider who may have brilliant ideas but who has rejected the system and has been rejected by the system. He will not be in a position to create new promotion paths or to protect young officers who avail themselves of them. Civilians, acting alone, are not entirely "legitimate" political players in the promotion of officers. The more professional a military organization is, the more it will collectively feel that only fellow professional officers are competent to select candidates for promotion. A civilian will have most impact if he can devise a strategy that reinforces the actions of senior officers who already have "legitimate" power in the military. Civilians are outsiders to the community; they can favor their military protégés, but they cannot make them "legitimate" in the eyes of the officer corps if they are not already seen as such. Civilian outsiders can, however, help protect military innovators against some internal and external opposition.

In short, this perspective suggests that peacetime military innovation occurs when respected senior military officers formulate a strategy for innovation, which has both intellectual and organizational components. Civilian intervention is effective to the extent that it can support or protect these officers. This perspective is not helpful in examining every peacetime military dispute. It does not address normal budget disputes within the military or between civilians and the military. It does not seek to explain the outcome of debates about program choices or short-term policy options that do not affect the fundamental values of the military. But it does help us understand more clearly a broad range of peacetime innovations, both successful and failed, in the modern American military.

The framework presented above also may not fully explain truly unique innovations, such as the introduction of nuclear weapons, but it helps us understand this case is anomalous. With nuclear weapons, the radical novelty of the new weapon was and is such that the officer corps does not possess a body of professional expertise that gives it a legitimate right to exclude civilians from decisions about how nuclear war should be fought and which theories should dominate operational behavior, innovation, and promotions. A familiar and heartfelt refrain in the United States in the 1940s and 1950s was that no one, military or civilian, had ever fought a nuclear war, and that military men and civilians were thus equally competent to grapple with these new problems. In fact, if anyone possessed special knowledge that gave them the right to make decisions about nuclear war, it initially appeared to be the civilian scientists who had invented the bomb.

In chapters 2 and 3, we will examine specific peacetime innovations in some detail in order to see how, with the exceptions noted above, this explanation squares with historical experience.

THE PROBLEM OF WARTIME INNOVATION

Having carefully developed a picture of the peacetime military and the likely character of innovation in it, it is disconcerting but entirely necessary to go back to the beginning in order to understand wartime military innovation. This is simply a recognition that in wartime, everything is different. During wartime military organizations are "in business." They have less the character of a stable political communities in which citizens argue about values and acquire power by following established pathways and more the character of a functioning bureaucracy that has the strongest possible incentives to learn rationally from its experiences. Wartime military organizations, however, are in a very peculiar kind of "business." They face an enemy who is trying to destroy them physically, to thwart them psychologically, and in particular, to stop them from learning useful lessons that can then be used against the enemy. Military organizations face time horizons that are different from those of peacetime. They are no longer operating in an environment in which there are decades, perhaps generations, before war will come and innovations will be tested in combat. They are, instead, in the midst of war, and innovations must be thought through and implemented within two or three years if they are to be of any use at all. And, most obviously and importantly, the organization is suffering casualties in battle, and the morale and composition of the leadership is affected by that reality.

The problem of wartime innovation thus becomes one of the extent to which organizational learning can take place under the unique conditions of war. While constraints exist, so do opportunities. Opportunities for innovation exist because both old and innovative methods can be tested in combat and compared, and because old military leaders may die in combat and be replaced by younger officers. Theories of organizational learning based on the study of organizations that do not face the malevolent, violent chaos of war may not be useful in studying learning and innovation during war. Therefore, rather than beginning with general theories of organizational behavior, as we did in the section on peacetime innovation, it may be more useful here to begin by comparing some "common sense" ideas about war and wartime innovation with actual experience.

The general idea that war provides the necessary environment for military learning and innovation is widespread. After observing the incoherent and often misguided ways in which the Royal Air Force and the U.S. Army Air Forces tried in the years before 1939 to innovate and prepare for war, the authors of the official British history of the strategic bombing campaign against Germany partially excused the peacetime errors of those two services by noting that "it is in war itself that men learn how to fight—if time is given to them in which to learn."[45] The authors of the U.S. Strategic Bombing Survey concluded that airpower was in its infancy before the war, and only reached "full adolescence" during the war.[46] Another historian observed the decision of the United States Army Air Forces to conduct daylight bombing missions of ball-bearing factories in Germany in August and October 1943 without long-range fighter escorts to protect the bombers all the way to their targets and back to base. Between one-fourth and one-third of the U.S. bombers were destroyed in those raids, and daylight missions in good weather were cut back until new long-range fighters were available in sufficient quantity as escorts. In retrospect, the historian concludes, "It had been a murderous but essential learning process. Nothing but such experience could teach the USAF and its masters from General 'Hap' Arnold . . . downwards, about what it could not—and often about what it could—do."[47] Wartime experience forced the U.S. Army Air

---

[45]Webster and Frankland, *SAOG*, 1:126.

[46]*The United States Strategic Bombing Survey, Summary Report (European War)*, vol. 1 (GPO), 30 September 1945, p. 1.

[47]Ronald Lewin, *Ultra Goes to War* (New York: McGraw-Hill, 1978), pp. 242, 248–49. Sixty bombers out of 315 attacking USAAF bombers had been shot down over Regensburg and Schweinfurt on 17 August 1943 and 60 shot down and 17 heavily damaged out of 228 attacking bombers over Schweinfurt on 14 October 1943.

Forces to adopt the long-range escort fighters.

Yet there are so many examples of military organizations that have been unable, for whatever reasons, to learn from wartime experience that we are forced to be cautious in assuming that innovation during wartime is a straightforward matter of observing what works and does not work in combat. The British attempt in April 1915 to land large numbers of troops at Gallipoli was an innovative effort in response to the deadlock in France and Belgium. The simple view of organizational learning would argue that the perspective was that infantry assaults in Europe "didn't work," and so a search for new methods was undertaken, which led to the development of the tank and also to the amphibious operation at Gallipoli. But there were ferocious arguments at the time about whether operations in France were "working" or not, as we shall see in chapter 4, because it was not clear how "working" should be defined.

Assuming what is not the case, that wartime failure is unambiguous, what would an acknowledged failure teach the organization about what should be done, and how it should be done, as opposed to what should be avoided? How could the British learn how to conduct large-scale amphibious assaults in Turkey from the failure of infantry attacks in France? In 1914, there was no settled idea within the General Staff about how to conduct this kind of operation, and no forces trained to perform it, so the landing was a genuine attempted innovation. Both the initial landings on 25 April 1915 and the follow-up landing on the night of 6–7 August showed how difficult learning and innovation in wartime can be.

If the troops landing on shore in daylight from boats were not to be annihilated by land-based rapid-fire rifles and artillery, those weapons would have to be neutralized by naval gunfire. What wartime lessons were available to teach the British how this should be done? The commander of the expedition, General Sir Ian Hamilton, recorded in his diary ten days before the first landing: "[The] principle of naval gunnery is different from the principles of garrison and field artillery shooting. Before they will be much good at landmarks, the sailors will have to take lessons in the art."[48] There had been no opportunity for the Royal Navy to learn how to shoot in support of amphibious operations instead of at ships. Nonetheless, it was assumed that at Cape Helles, where the land jutted out and seemed to put the Turkish positions at the mercy of the Royal Navy,

[48]Ian Hamilton, *Gallipoli Diary* 2 vols. (New York: George H. Doran, 1920), 1:113 (entry for 15 April 1915).

naval gunfire would create conditions suitable for the first set of landings. But when at 6:30 A.M. the naval guns ceased fire and the troops came ashore at Sedd-el-Bahr, Turkish riflemen opened up with murderous effect.[49] Hamilton had these observations. "The whole landing place at 'V' Beach is ringed round with fire. The shots from our naval guns, smashing as their impact appears, might as well be confetti for all the effect they have had upon the Turkish trenches. . . . And the bay at Sedd-el-Bahr, so the last messengers have told us, had turned red."[50]

A major innovation, is, by definition, unprecedented. Even if that innovation takes place in wartime, there will not have been much relevant previous experience. The lack of precedent makes wartime innovation risky, and with the risk often comes a justified aversion.

If the decision to try an innovation in wartime is made, surely it is reasonable to hope that its relative utility can be evaluated and that it can be used or modified thereafter. Thus, we might expect that at the very least the second landing on the Gallipoli Peninsula at Suvla Bay in August would have benefited from the experiences of the first landings. To some extent, it did. Logistical preparations were far more thorough, and landing craft were improved. But the lessons of combat are by no means unambiguous even when viewed first hand. Hamilton believed that the first landings demonstrated the importance of surprise and rapid movement once ashore to take advantage of the surprise. If given time to react, the defender with modern firepower would be able to resist a far larger offensive force.[51] However, General Stopford, operational commander of the Suvla landing, and his staff officer, Brigadier Reed, believed, on the basis of the experience in France, that is was foolhardy to assault enemy entrenchment without a thorough bombardment by onshore, as opposed to shipboard artillery. "All the teaching of the campaign in France", Stopford argued, "proves that continuous trenches cannot be attacked without the assistance of a large number of howitzers." Hamilton declined to overrule him, and on the day of invasion the landing force of twenty thousand men, who were to have advanced rapidly to cut the Turkish lines of communication before reinforcements could be brought down, allowed themselves to be held up by three thousand Turkish troops because no howitzers had been landed. By the time the British force began to advance, the

---

[49]Alan Moorehead, *Gallipoli* (London: Hamish Hamilton, 1967), pp. 140–41.
[50]Hamilton, *Gallipoli Diary*, 1:135, 1:138 (entry for 25 April 1915).
[51]Ibid., 2:2; Moorehead, *Gallipoli*, p. 245.

Turks had already taken up strong defensive positions encircling the landing zone, and they kept the British bottled up.[52] One distinguished observer wrote that the problem at Suvla was that wartime lessons had been learned all too well. The secretary of the British War Council, Maurice Hankey, wrote in his dispatch from Suvla: "It looks as though this accursed trench warfare in France had sunk so deep into our military system that all idea of the offensive had been killed."[53] Thus it was far from clear which lessons of wartime experience were relevant, or how these lessons might be applied in providing a basis for innovation.[54]

In generalizing from the experience of Gallipoli, we might suggest that the real lesson here is that in wartime organizational learning can take place only when there is continuity of command; if only Hamilton, and not Stopford, had been operational commander at Suvla, the lessons of the first landings would not have been wasted. Yet one of the most striking examples of an organizational failure to learn can be related to a continuity of command. A major failure to innovate resulted from the impossibility of firing a successful commander, even when he was unwilling to obey direct orders to change his modes of operation. In January 1942, Chief Air Marshall Arthur Harris took command of the Royal Air Force's Bomber Command; at that time it had been unsuccessful in damaging either military or industrial targets in Germany. Harris bolstered the command's morale and presided over its growth and the first mass attacks on German cities. The Bomber Command was then the only weapon Britain had for striking directly at Germany. Given the relatively poor accuracy shown in the command's night bombing raids in 1942 and his beliefs about the vulnerabilities of German society, Harris favored attacks on urban centers, as opposed to precision attacks on key industrial sites. Yet by mid-1944, the accuracy of British bombing had radically improved, and intelligence was available that suggested that attacks on cities had less value, and attacks on the German oil industry more, than Harris supposed. Direct orders were given Harris on 25 September 1944 to make the German oil industry the priority target of Bomber Command. Yet in October, only 6 percent of the tons of British bombs dropped on Germany

---

[52]Moorehead, *Gallipoli*, pp. 257, 261, 264–65, 291.

[53]Lord Hankey, *The Supreme Command 1914–1918* 2 vols. (London: George Allen and Unwin, 1961), 1:399.

[54]For an insightful analysis of the problems of the landing at Suvla see John Gooch, "Failure to Adapt: The British at Gallipoli, August 1915," in Eliot Cohen and John Gooch, *Military Misfortunes* (New York: Free Press, 1990).

were intended for oil-industry targets. Despite lengthy discussions with the chief of the Air Staff, Sir Charles Portal, Harris continued to refuse to accept the priority of oil targets. Because of Harris's popularity, Portal could not fire him, and he could not make him change his mind. In December, Portal gave in, writing to Harris that he hoped the commander would bomb oil-industry targets to the best of his ability.[55] Continuity of command can lock in place doctrine adopted early despite subsequent experience, blocking innovation after an initial period of experimentation.

Thus few easy generalizations about wartime learning and innovation seem plausible. But the basic question remains: Can wartime experience help an organization innovate and improve its wartime performance? The question becomes more manageable if a distinction is made between two kinds of changes in military organizational behavior in wartime. A good empirical case can be made for the position that organizational learning can occur in wartime if it takes place in the context of existing military missions. Routine military intelligence and "feedback" have helped wartime organizations to learn, to reform themselves, and to improve their ability to execute established missions. But such improvements are not innovative as defined here. When better performance of an existing mission only makes the strategic situation worse, when a new mission for a combat arm must be invented in order to achieve victory, organizational learning and innovation become extremely difficult. This distinction derives from the views of two theoreticians who have addressed the specific problem of learning in war and from one of the best developed theories of organizational learning and change, the cybernetic model of decisionmaking.

The two classic theoretical studies of military behavior in wartime, Sun Tzu's *The Art of War* and Carl von Clausewitz's *On War*, provide very different views on what can and cannot be learned in wartime. Both works deal with military organizations that did not have the highly developed bureaucratic structures for handling information characteristic of modern military organizations, but they nonetheless provide a useful starting point for an examination of wartime learning.

Clausewitz's view on the availability of information in wartime is summed up in the phrase "the fog of war." Clausewitz fully appre-

---

[55]Webster and Frankland, *SAOG*, 3:63, 3:67, 3:73–74, 3:84–94. See also Solly Zuckerman, "Strategic Bombing and the Defeat of Germany," *Journal of the Royal United Services Institution* (June 1985), 67–70.

ciated the value of good intelligence as the basis for action, and, indeed, he argued that a commander needed to have far more information than the standard kinds of military intelligence would provide. He insisted that the size of one's own forces, the degree of force to be applied, expectations as to the duration of the war, and the proper target of one's own military efforts, the famous "center of gravity," all must be derived from an understanding of the political character of the enemy government, enemy public opinion, the personalities of enemy leaders, the nature of enemy alliances, of other nations that may become involved in the war, and of oneself.[56] He was, however, extremely pessimistic about the practical possibility of obtaining this theoretically desirable information. Before the war starts, he asserted, information will be imperfect, and after it starts, matters will only get worse. During war, "the latest reports do not arrive all at once; they merely trickle in." Worse, "many intelligence reports in war are contradictory, even more are false, and most are uncertain. . . . [T]he effect of fear is to multiply lies and inaccuracies." As a result, he had little use for military theorists who demanded that strategy be made only on the basis of reliable intelligence.[57] While Clausewitz had great respect for the intellectual capacity of great military commanders, he tended to emphasize the need for resolution and strength of character in wartime commanders, as opposed to analytical capabilities, precisely because they would have to make decisions in the absence of good intelligence.

If these assumptions concerning the worthiness of wartime intelligence are granted, then it follows that the occasions on which wartime learning and innovation are possible would be exceptional. It would then be best to stick to prewar plans, not to be rattled by the false information that would come in during battle, and to put the supreme commander on the battlefield to make tactical adjustments to the plan as he saw fit.[58] As for knowing the enemy, Clausewitz warned that "war is not an exercise of the will directed against inanimate matter, [but] is directed against an animate object that reacts."[59] So even if accurate information were obtained about the enemy in wartime, it could not be assumed to remain accurate; the

---

[56]Carl von Clausewitz, *On War*, trans. and ed. Michael Howard and Peter Paret (Princeton: Princeton University Press, 1976), book 1, chap. 1 (hereinafter cited in the form I,1), p. 77; VIII, 3b, p. 385; VIII, 3, p. 386; VIII, 4, pp. 595–96.

[57]Ibid., I,3, p. 102; I,6, p. 117.

[58]Ibid., III, 1, pp. 177–78; I,6, p. 118.

[59]Ibid., II,3, p. 149.

enemy would, in fact, try to change his behavior in the fashion most inconvenient to you.

If Clausewitz's views on the difficulties of learning in wartime can be summed up by phrase "The fog of war," Sun Tzu's views on wartime intelligence are summed up by two of his own aphorisms: "All warfare is based on deception," and "Know the enemy and know yourself, in a hundred battles you will never be in peril."[60] Sun Tzu's *The Art of War* can be understood as set of theoretical and practical guidelines on how to obtain information about the enemy in wartime and how to deny him information about oneself. It is not an accident that *The Art of War* begins with a chapter titled "Assessments," and finishes with one called "Employment of Secret Agents." The successful conduct of war hinges on the ability to assess the enemy and to prevent him from learning about you. Intelligence and counterintelligence are the keys to military operations and should be the basis of wartime strategy. Consider Sun Tzu's injunctions: "Success depends on foreknowledge. What is called foreknowledge cannot be elicited from spirits, nor from gods, nor by analogy with past events, nor from calculations. It must be obtained from men who know the enemy situation." Only the wartime commanders "who are able to use the most intelligent people as agents are certain to achieve great things. Secret operations are essential in war."[61]

Sun Tzu gives just as much attention to the necessity of studying oneself, to showing how a commander must learn what is going on with his own forces, must be able to tell when subordinate commanders are responsible for bad troop morale, when the nerves of a commander are shot, when officers have the genuine loyalty of their troops.[62] For Sun Tzu, learning about the enemy during war is only one part of the assessment. Measuring one's own strengths and weaknesses is the other essential aspect.

How do these two very different perspectives on intelligence and learning during wartime relate to the problem of innovation in modern military organizations? An answer must begin with an examination of the ways in which military organizations collect and use information.

All military organizations have mechanisms for collecting information about the enemy in wartime and for monitoring the state of

[60]Sun Tzu, *The Art of War*, trans. Samuel B. Griffith (New York: Oxford University Press, 1971), book I, parag. 17 (I–17); III–31.
[61]Ibid., XIII–3, XIII–4.
[62]Ibid., X–9, X–13, X–17 through X–21.

[29]

their own internal conditions. Determining the number and location of enemy units and the number and effectiveness of one's own forces are the bread and butter of military intelligence and internal military administration, respectively. Are these mechanisms adequate for learning and improving organizational performance in wartime?

In important ways they are, but they are capable only in the context of established missions. Routine measures for keeping track of the enemy and of the condition of friendly troops can be understood in terms of feedback, a concept originally developed for cybernetics. Feedback is the use of narrowly focused categories of information about the state of a system that are used to control that system in order to keep it in conformance with a previously specified desired state. The best-known example is the thermostat, which monitors the temperature in a room and adjusts the furnace so as to keep the room temperature constant. This concept has been applied to the behavior of organizations in peacetime,[63] and it can be applied to some cases of organizational change best thought of as *reform*, a limited form of innovation that notes failures of an organization to reach agreed upon standards of performance and makes adjustments.

Military organizations often experience reverses or failures in battle and are compelled to change their behavior. When that failure is the result of an inability adequately to perform tasks that have been well defined and that continue to be accepted as legitimate by the organization, the necessary changes can be characterized as reform. While improving the performance of existing organizational tasks is extremely difficult in wartime, using routine sources of information merely to identify the need for reform is not organizationally difficult, since a failure to meet goals defined by existing missions and concepts of operation is highlighted by the normal functioning of the organization. The organization can then act to bring itself back into conformance with its desired state of being, thus closing the feedback loop.

Consider, for example, the conduct of the German army after the 1939 Polish campaign. After-action reports were reviewed by the German high command. The soundness of the tactical concepts, organization, and weapons with which the German army entered the war was confirmed by the campaign, but the overall level of combat

---

[63]John D. Steinbrunner, *The Cybernetic Theory of Decision Making: New Dimensions of Political Analysis* (Princeton: Princeton University Press, 1974), passim; Robert F. Coulam, *Illusions of Choice: The F-111 and the Problems of Weapons Acquisition Reform* (Princeton: Princeton University Press, 1977), pp. 17–18.

effectiveness fell short of expectations, in large part because the rapid expansion of the German army in the previous five years had resulted in newly created units that had not received adequate training. The Army knew what it had meant to do, and could note how far it had fallen short. It could then rectify the situation by imposing more intensive training along the lines of that already in use. The standard mechanisms of military management functioned as they were designed to function; they indicated shortcomings within a basically sound framework. Williamson Murray correctly analyzed this as a "case study in professionalism."[64] What is most remarkable is that these mechanisms functioned even after a striking victory, after which the natural tendency would have been to rest easy.

An example of how this kind of feedback can operate in wartime after a defeat is the adjustment made by the British and American in World War II armies to a new military environment: jungle warfare. In Burma and Malaya, the Japanese initially inflicted serious losses on the British. Beginning in 1942, British officers began the job of reconstructing their army. They had little preparation for jungle warfare, and the only British units that had received such training had been sent off to fight in the deserts of Iraq.[65] Standard U.S. Army divisions arrived in New Guinea and Burma and suffered defeat at the hands of the Japanese, after which they, too, had to rebuild.

What is striking about the period following the initial defeats is the straightforwardness of the adjustment to jungle warfare, how well the standard military organizations managed the problem. In adapting to the heat and disease, for example, it was discovered that the oldest and most traditional U.S. Army units did the best, because they had the tightest discipline. Orders by officers to take antimalarial medicines were strictly obeyed, whereas in newer units, the discipline was more lax.[66] Beyond this was the need to find out how the Japanese fought in order to develop responses. The methods employed were the standard tools of military intelligence— debriefing combat commanders to discover what had gone wrong or

[64]Williamson Murray, "The German Response to Victory in Poland: A Case Study in Professionalism," *Armed Forces and Society* 7 (Winter 1981), 288–91.

[65]F. Spencer Chapman, *The Jungle Is Neutral* (London: Chatto and Windus, 1949), p. 29; Field Marshall Sir William Slim, *Defeat into Victory* (London: Cassel, 1954), p. 20.

[66]Lincoln R. Thiesmeyer and John E. Burchard, *Science in World War II: Combat Scientists* (Boston: Little, Brown, 1947), p. 137. Despite its slightly lurid title, this is the excellent official history of the Office of Scientific Research and Development's Office of Field Service, which placed scientists and engineers in combat units to help with new technologies, including medical technologies.

[31]

right in battle and sending out reconnaissance patrols to keep track of the enemy. When General William Slim was first brought to Burma, he was given the opportunity to speak with the only Allied commander who had succeeded in battle: a Chinese general who had defeated the Japanese at Changsha.

> His experience was that the Japanese, confident in their own prowess, frequently attacked on a very small administrative margin of safety. He estimated that a Japanese force would usually not have more than nine days of supplies available. If you could hold the Japanese for that length of time, prevent them capturing your supplies, and then could counter-attack them, you would destroy them. I listened to him with interest. . . . There were, of course, certain snags in the application of this theory, but I thought its main principles sound. I remembered it, and, later, acted on it.[67]

British units had to regain their confidence and become much lighter in order to move off the roads. Slim himself demonstrated that he was one of the great field commanders of World War II by successfully rebuilding the British Army in India. But his account makes clear that the response to the Japanese involved the use of routine military intelligence to modify light infantry units to handle the particular style of the Japanese, not a fundamental change in practices.

The experience of the American units in New Guinea and Burma paralleled that of the British. The American commander of the forces fighting in New Guinea in 1942, General Robert Eichelberger, found the Japanese and the jungle difficult enemies that could, nonetheless be fought off by restoring the discipline and morale of defeated units. The U.S. Army's Thirty-second Division initially suffered because its men had poor trail discipline: they abandoned essential equipment while on the march and drank contaminated water. It became an effective jungle fighting force when Eichelberger relieved exhausted commanders and installed more aggressive officers in their place. Eichelberger himself was always at the front lines, and his example inspired his men. In this he exemplified the conventional wisdom of the U.S. Army's *Manual for Commanders:* "The morale of a unit is that of its leader. . . . The presence of a commander with the troops in action, as often as possible, is essential to morale." The Thirty-second Division managed the problem of jungle

[67]Slim, *Defeat into Victory*, p. 18.

fighting when it was given a commander who had the strength and intelligence to apply standard procedures to a difficult situation.[68] Similarly, General Joseph Stilwell had remarkable success in Burma in building three Chinese divisions, units that were able to defeat the Japanese using standard American division organization and training methods.[69]

Other examples of military reform in wartime are not hard to find. General Matthew Ridgway assumed command of the badly beaten and demoralized Eighth Army in Korea in 1951. He, too, faced a new problem, the unconventional army of the People's Republic of China, which utilized lightly armed troops who were masters of infiltration, deception, and concealment. Yet he was able to rebuild his forces into a victorious army by appealing to the most traditional of U.S. Army doctrines—indeed by telling his officers that their infantry forefathers would roll over in their graves if they saw how Eighth Army officers had failed to seize the high ground, had not taken prisoners to determine enemy strength and positions, and had not ensured that infantrymen attended to such basic things as maintaining contact with friendly units on their flanks, wearing helmets, and keeping full loads of ammunition. When some of his commanders complained that their radios would not work in the mountainous terrain of Korea, he told them to send out runners on foot to maintain communication, as their grandfathers had done, or to even send up smoke signals.[70]

In each of these cases, the performance had fallen short of goals defined by the traditional tasks of the infantry. The Japanese had used the jungle to outflank the British and the Americans; the Chinese had used deception and camouflage to exploit lapses in American military methods. Rectifying these situations required heroic effort, but both the problems and the solutions were matters of established organizational routine. The need to be mobile was and is a traditional infantry performance goal. If the Japanese were

---

[68]Robert L. Eichelberger, *Our Jungle Road to Tokyo* (Washington, D.C.: Zenger, 1949, rp. 1982), pp. 11–14; Jay Luvaas, "Buna, 19 November 1942–2 January 1943: A Leavenworth Nightmare," in Charles E. Heller and William A. Stoft, eds., *America's First Battles, 1776–1965* (Lawrence: University Press of Kansas, 1986), pp. 197, 211–12, 221–24.

[69]Charles F. Romanus and Riley Sunderland, *Stilwell's Mission to China* (Washington, D.C.: Department of the Army, GPO, 1953), pp. 212–19; Romanus and Sunderland, *Stilwell's Command Problems* (Washington, D.C.: Department of the Army, GPO, 1956), pp. 27–28.

[70]Matthew B. Ridgway, *The Korean War* (New York: Da Capo Press, 1967), pp. 88–89; S. L. A. Marshall, *The River and the Gauntlet* (New York: Time Book Division, 1953), pp. 18, 54.

more mobile than the British through the jungle, the British had to become more mobile. Taking advantage of terrain is another traditional infantry objective. If the Chinese in Korea were making better use of terrain than the Americans, than the Americans had to improve. Although the Japanese and Chinese presented traditional military units with new problems, those problems could be handled by the routine operation of organizational feedback, in which established categories of data indicate deviations from the preferred state of affairs.

But what happens when a new wartime problem occurs that falls outside the parameters of established missions and concepts of operation? What if the pursuit of existing performance goals only makes the problem worse, and a solution requires the development of completely new capabilities? It may not be reasonable to assume that the organization's existing mechanisms for collection will take in information that falls outside established categories. In wartime, this is a particularly relevant problem, because, by definition, an innovation will involve new organizational tasks and concepts of operation. Mechanisms set up to monitor existing functions may show that they are not being adequately performed, but will not collect information that lays the basis for the development of new functions. A thermostat has the overall goal of keeping a room's temperature comfortable, but it will not provide any data about the need to dehumidify air in a room. So, too, existing military intelligence and administrative routines may not suggest the need for or value of innovation.

An understanding of this problem demands an examination of how military organizations routinely measure their performance. Military missions are undertaken in pursuit of a strategic goal. In a major conventional war, this might be the destruction of the enemy army or battlefleet, command of the sea lines of communication, or any other such goal. Implicit is an understanding of how day-to-day military operations are related to that strategic goal. Defeating enemy ground forces in battle and seizing key points of terrain are related to winning larger campaigns, which are, in turn, linked to the destruction of the enemy army. Sinking enemy ships, seizing key ports, and establishing patrolled areas that friendly unarmed ships can safely use are linked to winning control of sea lines of communication. These relationships define measures that determine whether the military organization is performing the way it is supposed to perform. Intelligence about the location and number of friendly and enemy troops or ships in a given sector, that is to say the relative order of battle, and the quality of those forces, are operational measures of how the war is going. Taken together, the definition of the strategic goal, the relationship of military operations to that goal, and indicators of how well operations are proceeding can be thought of as a *strategic measure of effectiveness* for the military organization. This differs from operational measures of effectiveness. Armed forces may perform very well at the operational level, but because those operations are not linked to the strategic goal, they may be futile. The best known case is exemplified by what a U.S. Army officer told his North Vietnamese counterpart in 1975—that the United States Army had never been defeated in combat in South Vietnam. The North Vietnamese officer replied that that was true, but irrelevant, and he was correct.[71]

If an appropriate strategic measure of effectiveness is in place, information can be collected that is relevant to that measure, so that organizational learning leading to reform can take place. When military innovation is required in wartime, however, *it is because an inappropriate strategic goal is being pursued, or because the relationship between military operations and that goal has been misunderstood.* The old ways of war are employed, but no matter how well, the war is not being won. A new strategic goal must be selected and a new relationship between military operations and that goal must be defined. Until that happens, information will be collected that is relevant to the old goals and relations, and there is no reason to suppose that this information will suggest new, alternative ways of winning the war. Until the strategic measure of effectiveness has been redefined, organizational learning relevant to innovation cannot take place. When a new goal has been defined, and a new relationship between military operations and victory established, then, and only then, can information be collected and analyzed that will determine whether the old ways of war are better or worse than a proposed set of new alternative ways of fighting. A redefinition of the strategic measure of effectiveness tells the organization what and how it should be learning from wartime experiences. Until such a redefinition takes place, a wartime military organization will learn from its experiences in terms of existing measures. Failure to do well according to the old measure of effectiveness will only prompt renewed effort to become more effective in ways so defined. Inappropriate strategic measures of effectiveness may lead an organization to mistakenly

[71]Harry G. Summers, *On Strategy: A Critical Analysis of the Vietnam War* (Novato, Calif.: Presidio Press, 1982), p. 1.

increase its efforts, in a vicious circle, at a time when increasing the effort put into old methods only draws the organization deeper into failure. The thermostat can "learn" whether it has allowed the furnace to run for too long or short a time, but until it is given a new goal of controlling humidity, and a device to measure humidity is added, it will never learn that a change of air is needed.

The suspicion that innovative learning will not occur in wartime without a new strategic measure of effectiveness is reinforced by looking at "natural experiments" in which military organizations failed to innovate in wartime. In the case of the British Navy against German U-boats in World War I, old measures of effectiveness only led the British into redoubling their ineffective antisubmarine patrols. This contributed to the delays in instituting convoys to protect Allied shipping. The declaration of unrestricted submarine warfare in 1917 presented the Royal Navy and the British nation with the prospect of defeat by starvation. In the first three months of 1917, more merchant ships were lost to German submarines than had been constructed by Great Britain thus far in the war. The reaction of the Admiralty was to redouble its demands for active patrolling of the areas in which submarines were thought to be operating, patrols that had not provided the necessary protection for merchant ships. The feedback loop in which more submarine attacks produced more antisubmarine patrols only reinforced failure. Suggestions that the Admiralty turn its attention instead to providing escorts to merchant ship convoys were consistently rejected, despite the successful precedents of escorted convoys to France and to and from Scandinavia and despite the endorsement of convoys in 1917 by some Royal Naval officers including the commander in chief of the fleet, Admiral Beatty, and the senior Royal Navy commanders in the Mediterranean.

One of the principal reasons for rejecting convoys given by the Admiralty was that there were not enough warships to convoy all the transatlantic shipping entering and leaving British ports. Because convoying had been rejected before World War I, it was not an established mission, and there were no mechanisms for monitoring the number of ships that made transatlantic voyages to and from British ports. The Royal Navy relied upon the Customs Office for counts of the number of ships entering and leaving British ports. Unfortunately, Customs Office data counted all ships entering and leaving British ports, including small vessels engaged in coastal trade. The figure was 5,000 ships clearing British ports each week, far too many to be convoyed. It was not until a Royal Navy commander in charge of the British convoys to France reviewed the Customs Office figures in the spring of 1917 that it was realized the the number of large ocean-going merchant ships entering and leaving British ports amounted to only 120 to 140 a week, a number that could easily be convoyed with Royal Navy resources. The historian Arthur Marder notes that the figure of 5,000 clearings a week may not have been taken entirely seriously inside the Royal Navy and that the opposition to convoys had other sources, but he clearly faults the Royal Navy for not developing adequate sources of information. The prime minister, David Lloyd George, emphasized this analytical failure in condemning the Royal Navy's delay in instituting convoys.[72]

Wartime innovation was impeded in this case by the use of existing but inappropriate measures of performance. The Royal Navy's resistance to innovation had its origin in a commitment to offensive methods of antisubmarine patrolling. What, in retrospect, we see as the obvious failure of these practices was not enough to overcome this resistance, because the feedback loop only saw the need to increase the level of existing activities, and because the data relevant to innovation were not being utilized. For the resistance to be overcome the Royal Navy had to learn to look at different categories of information. Instead of focusing on enemy submarines sunk by patrols, it had to focus on the number of merchant ships at risk to submarines. The new strategic measure of effectiveness was the proportion of transatlantic sailings that survived, a proportion the Navy could not determine before the spring of 1917 because its statistical measures were oriented toward other goals. Eliot Cohen has noted that a similar misdefinition of the requirements of organizational learning led initially to inappropriate responses on the part of the U.S. Navy when it was first confronted with the problem of handling German U-boat attacks on U.S. merchant shipping in the Atlantic in 1942.[73]

The failure of the Royal Air Force to change the character of its strategic bombing is another case of a failure to innovate in wartime because of the absence of new, more appropriate strategic measures of effectiveness. Solly Zuckerman has argued that this lack of innovation sprang directly from the RAF's failure to redefine its strategic measure of effectiveness. By 1942, the dominant strategic goal of

[72]Arthur J. Marder, *From the Dreadnought to Scapa Flow: The Royal Navy in the Fisher Era 1904–1919*, 5 vols. (London: Oxford University Press, 1969), 4:64, 4:102, 4:118–19, 4:139–142, 4:144–45, 4:150–52.
[73]"Failure to Learn: American Antisubmarine Warfare in 1942," in Cohen and Gooch, *Military Misfortunes*, pp. 87–88.

the Royal Air Force's Bomber Command was to destroy the enemy work force's will to fight, which was to be achieved by area bombing of urban targets. Progress could be measured day to day by photographing and measuring the area of burned out urban areas and calculating the number of German workers "de-housed."[74] This strategy may have been appropriate early in the war, but as bombing accuracy improved, alternative strategies became possible. Zuckerman argues that policy did not change because the senior RAF leadership failed "to accept the notion that [road and rail] communications are the common denominator which affects all possible target systems." The strategic goal should have been shifted from undermining civilian morale to destroying the ability of the enemy to move his forces and industrial supplies. Because this redefinition was not effected, data that showed the potential impact of systematic attacks of bombing on transportation systems were overlooked. Organizational learning to support innovation did not take place.[75]

This somewhat bewildering review of wartime lessons learned and not learned suggests that, as far as the descriptive side of wartime innovation is concerned, Clausewitz may have been more correct then Sun Tzu. Intelligence relevant to innovation very likely will not be available in wartime, and wartime innovation is likely to be limited in its impact. Consider the obstacles to wartime innovation. First, the collection of intelligence that fits into any existing strategic categories will be difficult, for all the reasons noted by Clausewitz. In addition, new measures of strategic effectiveness must be invented, new methods of intelligence collection developed, and successful organizational innovations developed in response to that intelligence, all within the few years of active fighting. These circumstances suggest that wartime innovation will be limited in its impact where it does occur at all, because the time necessary to complete all these tasks is likely to be long relative to the length of the war.

There are two implications of this observation. The more obvious is that the longer the war lasts, the greater the impact of wartime innovation. Less obvious and more interesting is the idea that more hierarchical and centralized the organization, the greater the impact of the innovation is likely to be, because a tightly controlled organization in which intelligence is collected and concepts of operation

[74]See, for example, Max Hasting, *Bomber Command* (New York: Dial Press, 1979), pp. 131–32.
[75]Zuckerman, "Strategic Bombing," p. 79.

are enforced from the center may be able to act more quickly. It will have the ability to assemble a clear picture of the need for the innovation and to implement it in time for it to have some impact. In terms of amenability to innovation during wartime, the trade-offs between a loose organization that lets ideas circulate and a tight one that makes things happen quickly once an order is given seems to favor organizational tightness. The consequences for wartime innovation or organizational looseness will be explored in chapter 4 in a discussion on the invention of the tank in World War I.

What about the very plausible but contrary hypothesis that a decentralized organization will innovate more rapidly in wartime, because the men on the firing line will learn of the need for innovation first, and given the ability to act independently, will innovate more quickly than if they had to wait for orders from above? Decentralization would seem to favor innovation in those circumstances in which the operating units can collect all the relevant data themselves and can execute the innovation without the need for organizational changes elsewhere in their service. This kind of organizational structure will be explored in chapter 5 in relation to U.S. submarine warfare in the Pacific in World War II as it reshaped itself to fight an unrestricted war of a kind that had been forbidden it before December 7, 1941.

Finally, the differences between Clausewitz and Sun Tzu concerning the availability of intelligence in wartime will be examined in chapter 6 in relation to a case in which circumstances seemed to present the intelligence collector and innovator with the most favorable circumstances. While the U.S. Army Air Forces in World War II had to invent many key strategic measures of effectiveness during the war, it also had early access to good intelligence about the German air force and the effects of strategic bombing, because British intelligence had been able to intercept and decode German radio communications. What kind of innovation was possible under these circumstances, and how did it compare in effectiveness with peacetime innovation?

TECHNOLOGICAL INNOVATION

In examining technological innovation in the military, we are again drawn back analytically to square one. Peacetime and wartime organizational innovation, as defined in this book, is concerned with social innovation, with changing the way men and women in orga-

[39]

nizations behave. Technological innovation is concerned with building machines. Yet technological innovation in the military has an important political component as well. Technological innovation introduces a new dimension to the relationship between one's own forces and the military organization of the enemy, a qualitative technological one. It introduces a new set of domestic actors, scientists, into the community within which military decisions are made. In short, technological innovation gives rise to an additional set of questions beyond those associated with organizational innovation.

First, there is the simple empirical question. What has been known about the character of enemy military technology? Has information about enemy technology been the spur to innovation? Second, how can an organization evaluate the costs and benefits of weapons that do not yet exist? In other words, how can decisions about military research and development be made when the weapon to be produced does not represent an incremental change, but a radical break from the past? A related question is whether it is scientists or military men who are best qualified to make these decisions.

Technological innovation has long been the object of study in the nonmilitary world, where it has also been distinguished from social innovation. Simon Kuznets, for example, wrote that it was reasonable to study separately

> technical inventions yielding new products to be turned out and new devices to be used in economic production. We thus exclude social inventions, new methods of inducing human beings to compete and cooperate in social progress. . . . The effects of social inventions on economic productivity are obviously major and profound, but the occupational groups connected with social inventions and the institutional arrangements for their production, selection and application are so different from those involved with technical inventions that the two can hardly be treated together.[76]

The study of technological innovation in the government and business world, however, has not proceeded so far as to provide any clear-cut models for the study of technological innovation in the military. One pioneering work of the 1950s investigated the sources of

[76]Simon Kuznets, "Inventive Activity: Problems of Definition and Measurement," in *The Rate and Direction of Inventive Activity—Economic and Social Factors: A Conference of the Universities–National Bureau Committee for Economic Research and the Committee on Economic Growth for the Social Science Research Council* (Princeton: Princeton University Press, 1962), p. 19.

innovation in industry and asked such basic questions as whether there was an "optimal" rate of innovation, and if so, how would it be defined? Was the rate of innovation increasing with time? Was there any necessary connection between advances in basic science and technological advances? Were resources better used to fund many small research projects or a few major activities? The authors of the study did not provide any definitive answers, commenting, "One simple proof of the rudimentary state of knowledge is that practically any statement can be made without fear of decisive contradiction."[77] Other surveys of the literature on technological innovation in the business world asked if it was possible to determine whether scientific invention "pushed" technological development of new industrial products or processes, or rather whether social demand "pulled" technological innovations from inventors. Not surprisingly, the literature provided support for both these.[78]

The 1960s saw several efforts to do aggregated quantitative analysis of technological innovation in the business world, measured in terms of numbers of patents issued in a given sector, to determine both the sources of the innovations and their utility. One of the most thorough studies found that in the American railroad industry from 1839 to 1950 the number of patents issued and the level of investment moved together, that is, that when investment rose, so did the number of patents issued, and visa versa. But it also found that reversals in investment trends generally preceded reversals in patent trends by from one to three years. From this it was concluded that "the introduction and diffusion of inventions in the industry seem merely to accompany but not to cause the waves of investment." Innovation did not cause investment, but responded to it. Shifts in investment, it was then speculated, took place in response to "changes in the demand for railroad service." In this case, at least, it appeared that after the primary invention of the railroad, social demand and not scientific push tended to be source of innovation. The available data did not permit any answer to the question of whether the same pattern could be found in the agricultural, petroleum, or paper-making sectors of the American economy.[79] Other studies

---

[77]John Jewkes, David Sawers, Richard Stillerman, *The Sources of Invention* (London: Macmillan, 1958), p. 5–7.

[78]Richard R. Nelson, *Invention, Research and Development: A Survey of the Literature,* RM-2146 (Santa Monica, Calif.: Rand Corporation, 15 April 1958), pp. iii, 7–12.

[79]Jacob Schmookler, "Changes in Industry and in the State of Knowledge as Determinants of Industrial Innovation," in *Rate and Direction*, p. 201–2, 208–15.

[41]

found the number of patents issued in an industry to be correlated with improvements in productivity.[80]

Studies of particular industries that followed also tended to confirm the picture of demand pulling technological innovation. An international comparison of the genesis of electronic digital data processing equipment, that is, of the modern computer, found that in the 1930s and 1940s both Germans and Americans involved in endeavors that required the solution of large numbers of simultaneous linear or nonlinear differential equations found themselves facing mathematical problems that could not reliably be solved by human computation in reasonable periods of time. Konrad Zuse in Germany and Howard Aiken in the United States went on to develop primitive programmable digital computers. The application of electronic tubes to the problem of solving differential equations by digital means that produced ENIAC computer was driven by the wartime need to develop ballistic tables to predict artillery shell trajectories.[81] Most recently, a survey of industries in the U.S. found that patterns of innovation and their sources could be determined by looking at the expected short-term profits that were expected from an innovation. In this sense, expectation of profit also served to pull innovation along.[82]

But the relative role of demand in spurring innovation is clearly more complicated in the field of military technological innovation. One extensive survey of American weapons acquisition concluded that basic military missions had remained constant over time, and that changes therein could not therefore be understood as the source of technological innovation, which had to be found instead in new ideas generated in universities, research labs, and industry.[83] On the other hand, a U.S. Department of Defense (DOD) study, Project Hindsight, reviewed the history of twenty major U.S. weapons programs. It identified 710 discrete events in the history of the development of these weapons and found that "a clear understanding of a DOD need motivated 95 percent of all events (73% of all science

---

[80]Jora R. Minasian, "The Economics of Research and Development," in *Rate and Direction*, p. 140.

[81]Paul E. Cerruzi, *The Reckoners: The Prehistory of the Digital Computer, from Relays to the Stored Program Concepts, 1939–1945* (Westport, Conn., Greenwood Press, 1983), pp. 10–42, 43–44, 74, 107–9.

[82]Eric von Hippel, *The Sources of Innovation* (New York: Oxford University Press, 1988), pp. 4–5.

[83]Merton Peck and Frederick Scherer, *The Weapons Acquisition Process: An Economic Analysis* (Cambridge: Harvard University Press, 1962), p. 236.

events, and 97% of all technology events.)"[84] Yet Martin van Creveld surveyed modern military research and development and came unequivocally to the conclusion that military demand had not been the source of new military technologies: "During the twentieth century . . . none of the important devices that have transformed war—from the airplane through the tank, the jet engine, radar, the helicopter, the atom bomb and so on all the way down to the electronic computer—owed its origins to a doctrinal requirement laid down by people in uniform."[85]

In the case of the development of the jet engine in Great Britain and Germany in the 1930s, there was simultaneous "technology push" and "demand pull." The Royal Air Force displayed no interest in the early work of an individual RAF officer, Frank Whittle, on turbojets, and Whittle's own hopes lay with the civilian sector—he hoped to create market demand by introducing jet-powered aircraft for rapid mail delivery. At the same time, Hans von Ohain, who was working on turbine technology, was approached by the Heinkel aircraft company to help them develop more powerful engines that could take advantage of the stronger air frames being designed. In Germany, development of conventional piston-engines for aircraft lagged behind the rest of the world, and this drove Heinkel, as well as the Junkers aircraft company, to support work on jet engines.[86]

In the field of nuclear energy there is the obvious case of the development of the atomic bomb in the United States, which was clearly the result of scientific advances, and not of any preexisting demand,[87] and the equally clear case of the American submarine community searching among many technological alternatives that would permit a "true" submarine that could operate indefinitely underwater and creating the demand for the first nuclear reactors to produce mechanical or electrical rather than explosive power.[88]

---

[84]Raymond Isenson, "Project Hindsight: An Empirical Study of the Sources of Ideas Utilized in Operational Weapons Systems," in William H. Gruber and Donald G. Marquis, eds., *Factors in the Transfer of Technology* (Cambridge: MIT Press, 1969), p. 168. See also the summary of Project Hindsight findings in the *Department of Defense Annual Report for Fiscal Year 1967* (Washington, D.C.: GPO, 1969), p. 70.

[85]Martin van Creveld, *Technology and War: From 2000 B.C. to the Present* (New York: Free Press, 1989), p. 220.

[86]Robert L. Perry, *Innovation and Military Requirements: A Comparative Study*, RM-5182-PR (Santa Monica, Calif.: Rand Corporation, August 1967), pp. 15–18, 22–27.

[87]The story of the origins of this innovation are best laid out in Richard Rhodes, *The Making of the Atomic Bomb* (New York: Simon and Schuster, 1986).

[88]Hewlett and Duncan, *Nuclear Navy 1946–1962*, pp. 27, 41, 56–58, 68–72, 79–85.

Thus, whatever the case may be in industrial technology, the phenomenon of military technological innovation is not easily described in terms of "demand pull" or "technology push." A potentially more fruitful approach, suggested by the study of arms races, would be to examine the role that military intelligence plays in shaping technological innovation. Economic analyses of military research and development (R&D) have implicitly highlighted the role of intelligence in innovation. Deciding among alternative R&D proposals for new weapons could be considered the equivalent to a business investment decision, in which the present value of alternative investments can be compared by assessing their perceived value in the future and then discounting that by the appropriate interest rate to arrive at their present value. In the case of new weapons, however, both perceived future value and the relative urgency with which a weapon is needed (which would affect the implicit discount rate and present value—how much we are willing to spend today on an expensive new weapon that will not be ready for some time depends on when we think we will need to use it) are heavily influenced by intelligence about the enemy. Economically rational research and development decisions, as a result, are heavily dependent on intelligence that may be easy or difficult to obtain.[89]

The study of arms races thus suggests technological innovation in the military is or ought to be closely related to reliable intelligence about what the enemy is doing now and will be doing in the future. Studies of battleship design at the beginning of the twentieth century indicate that engineers of the industrialized nations had remarkably good access to information about each other's designs, and this information led to competitive innovation. German design philosophy, for example, was revolutionized by the need to respond to the British *Dreadnought* battleship.[90] American naval intelligence in the 1930s routinely collected information about foreign naval vessels, including displacement, maximum and cruising speeds, horsepower, range, armor, and firepower. The speed of the *North Carolina* class of American battleships was increased in the late 1930s when intelligence revealed that new Japanese battleships had reached higher speeds during their trial runs than had been

[89]Perry, *Innovation and Military Requirements*, pp. 6–9.

[90]Charles H. Fairbanks, Jr., "Choosing among Technologies in the Anglo-German Naval Arms Competition, 1898–1915," in William B. Cogar, ed., *Naval History: The Seventh Symposium of the U.S. Naval Academy* (Wilmington, Del.: Scholarly Resources, 1988), pp. 132–38.

expected.[91] Samuel P. Huntington has suggested that, in modern times, information about the number and quality of enemy forces has been more generally available than in the past, and that "the arms race in which the military preparations of two states are intimately and directly interested is . . . a modern phenomenon." Given the fact that in advanced nations the technological accomplishments of one scientific team can quickly be matched by another, innovations in one nation will trigger matching or responsive innovations in another.[92] The link between the activities in the competing nations is the network of military technical intelligence.

There are, however, examples in which important military technologies went unnoticed by the intelligence services of their enemies. The Japanese Zero fighter had twice the range of competing fighters, an important factor given the expansiveness of the Pacific region. This capability went unnoticed until a Zero was shot down over China and its wreckage analyzed in 1941. Even then, this knowledge did not find its way back to American or British aircraft designers, or to operational RAF or U.S. Air Force commands. New American aircraft comparable to the Zero had already been designed and built before detailed knowledge of the Zero was circulated to American aircraft manufacturers.[93] Knowledge about enemy military technology is one kind of intelligence relevant to our choices concerning military R&D, but intelligence about how the enemy will conduct himself is also important. As Huntington points out, however, such information may be difficult to obtain. There may be conflicting estimates of enemy capabilities, and misconceptions about the enemy may persist for long periods of time. There is, therefore, an important factual question to be answered. In specific cases, what was known of enemy behavior by the organizations responsible for military R&D? This question will be addressed in chapter 7 below.

The second major problem characterizing military technological innovation had to do not with the enemy, but with evaluating the

[91]"Study of Building Programs," Naval War College, August 1935, Record Group 8, Historical Collections, Naval War College, Newport, R.I.; Edwin P. Layton, Roger Pineau, and John Costello, *'And I Was There': Pearl Harbor and Midway—Breaking the Secrets* (New York: William Morrow, 1985), pp. 57–58.

[92]Samuel P. Huntington, "Arms Races: Prerequisites and Results," reprinted in Art and Waltz, eds., *The Use of Force*, pp. 366, 375, 392.

[93]Michael I. Handel, "Technological Surprise in War," *Intelligence and National Security* 2 (January 1987), 7–9; Allan M. Lazarus, "The Hellcat-Zero Myth," *Naval History* (Summer 1989), 49–50.

costs and benefits of new weapons. Perhaps there is possible some unambiguous measure of the value of the "output" of military technological innovation, such as "military power." If such a measure existed, then past military technological innovations could be compared with their costs. This would provide the basis for choosing among alternative innovations. Unfortunately, there does not seem to be any such unambiguous measure of the output of military technological innovation. One view expressed by experts is that the value of American military technology programs after World War II was very high. The economist Charles Hitch, looking back on this period, commented:

> By whatever arbitrary index of military power one measures the productivity of this [U.S. weapons] industry, it has increased since World War II not by a few percentage points per annum but by many orders of magnitude. And it has increased not because of a larger allocation of resources or their more efficient use within a given technological state-of-the-art constraint, but because of technological progress stemming from a very deliberate and expensive research and development program.[94]

But other academic observers of the same U.S. military research and development programs have questioned whether those technological innovations were useful at all, even apart from any question of cost. The Soviet technological establishment, it was argued, rapidly incorporated U.S. technological advances into Soviet military systems, leaving the United States with at best a fleeting technological advantage. At worst, innovation moved both parties into technologies that were in some fashion "destabilizing."[95]

A more systematic effort to judge the utility of U.S. military technological decisions suggests that, in the aggregate, when U.S. spending on military research and development exceeded Soviet spending, the U.S. lead in military relevant technologies increased. These leads did not emerge immediately when more money was spent, but after a lag of approximately five years, during which, presumably, research and development funding yielded results. The evidence for

[94]Commentary by Charles J. Hitch in *The Rate and Direction of Inventive Activity*, p. 194.

[95]Franklin Long and Judith Reppy, "Decision Making in Military R&D: An Introductory Overview," in Franklin Long and Judith Reppy, eds., *The Genesis of New Weapons: Decision Making for Military R&D* (New York: Pergammon Press, 1980), pp. 4–5.

this assertion is the relationship between Department of Defense spending on research and development and estimates of the relative military technological capabilities of the United States and the Soviet Union made by independent analysts and the U.S. government. Starting from a low level of spending in 1950, U.S. military research and development funding increased to approximately $12 billion (in 1980 dollars) in 1956, and rose to almost $20 billion (again in 1980 dollars) a year in the first half of the 1960s. Defense Department funding for research and development dropped after 1967, reflecting in part a conscious decision on the part of the McNamara administration that "with the United States having acquired a substantial stock of reliable and survivable strategic missiles on land and sea, further research in this area can probably be carried out at a more modest level."[96] Spending bottomed out in 1977, and beginning in 1978 the trend reversed, and R&D spending increased. Soviet spending on military research and development in dollars is estimated by the CIA. It is a very rough estimate, involving many unknowns and uncertainties, and can be used only to judge the most general trends. The estimates are that Soviet military research and development spending was perhaps one-fourth to one-third that of the United States in 1956 and increased at a steady rate until it overtook the United States between 1968 and 1972. U.S. funding began to approach Soviet levels by 1985.[97]

The levels of spending by the Defense Department on research and development are of course reflected in the human resources devoted to defense related research and engineering. Measured in terms of full-time equivalents of scientists and engineers, there were 114,000 men and women employed in defense-related research and development, including the university and industrial sectors, in 1954. That number climbed to 184,000 in 1958 and 209,000 in 1961,

---

[96]*Department of Defense Annual Report for the Fiscal Year 1964* (Washington, D.C., GPO 1966), p. 36.

[97]Figures on U.S. and S.U. R&D spending interpolated from graphs in *The FY 1980 Department of Defense Program for Research, Development, and Acquisition, Statement by the Honorable William J. Perry, Under Secretary of Defense, Research and Engineering to the Congress of the United States*, 96th Congress (Washington, D.C.: GPO, February 1979), p. 11–19 (hereinafter this and similar statements by the under secretaries of defense to the Congress will be referred to as *USDR&E Posture Statement*, followed by the fiscal year which it covers); and "Allocations of Resources in the Soviet Union and China—1985," Hearing before the Subcommittee on Economic Resources, Competitiveness, and Security Economics of the Joint Economic Committee of the United States, 99th Congress (Washington, D.C.: GPO, 1986), 2d sess., part 2, March 19, 1986, pp. 108–9, explanatory text pp. 104–5.

the peak year. The numbers declined to 160,000 in 1965, 188,000 in 1969, and 149,000 in 1971.[98]

In the United States, the level of spending on research and development was associated with the relative military technological standing of the United States and the Soviet Union, as measured in various ways. In the period 1945–1950, for example, before the Korean War increased the level of funding for all American military programs, the Soviet Union introduced thirty-five technological innovations in fighter and attack aircraft, versus twenty-three for the United States. This trend reversed after 1950, and in the period 1950–1955, the United Stated introduced eighteen tactical aviation innovations, and the Soviet Union thirteen. The performance of U.S. and Soviet jet engines, which can be measured with some objectivity, was such that a clear technical advantage for the United States did not emerge until the second half of the 1950s, reaching a lead, measured in the difference between dates on which comparable engines were introduced, of four to six years by 1967.[99] The timing of the emergence of this technological lead is broadly consistent with the turnaround in U.S. research and development funding starting in 1951.

The downturn in U.S. funding began in the late 1960s while Soviet spending on research and development continued its steady growth. The result was a closing of this technological gap between the two countries by the middle 1970s. Measuring the trends in this area is difficult. Directors, and then under secretaries, of defense for research, development, and engineering (USDR&E) began publishing charts comparing the levels of U.S. and Soviet technologies in various fields. These officials, unfortunately, did not choose to assess the same technology areas every year. This makes it difficult to track trends in relative technological performance. A rough assessment can be attempted, however. The USDR&E Posture Statement for FY 1977 included technology areas that were embodied in operational military systems, such as artillery technologies, as well as basic technologies, such as high pressure physics, computers, and

[98]National Science Foundation, *National Patterns of R&D Resources, 1953–1972*, NSF-72–300, December 1971, p. 34; cited in Frederick Seitz and Rodney Nichols, *Research and Development and the Prospects for International Security* (New York: National Strategy Information Center, Crane, Rusack, 1973), p. 57.

[99]Robert Perry, *Comparisons of Soviet and U.S. Technology*, R-827-PR (Santa Monica, Calif.: Rand Corporation, June 1973), pp. 19, 29, 31, 32.

high-energy lasers. In the assessment of twenty-one technology areas, the United States led in ten, the Soviet Union in eight, with parity or unknowns in three.[100]

In subsequent years, separate charts were compiled for basic technologies and technologies in deployed military systems. By the beginning of 1979, the official judgment was that in basic technologies the United States had a lead in twelve out of twenty areas and the Soviet Union in none. In deployed military systems, ranging from strategic weapons to ground force systems and naval systems, the United States or NATO led in nine out of twenty areas, the Soviet Union in eight, with three areas mixed or equal.[101] By 1986, the assessment was that in basic technologies, the United States led in fourteen out of twenty areas, and the Soviets in none. In deployed military systems, the United Stated led in sixteen out of thirty-one areas, the Soviets in six, including two areas (strategic bomber and ballistic missile defenses) in which the U.S. had no deployed systems to compare with Soviet systems, with parity in nine areas.[102]

The figures can be aggregated to determine the percentage of the technology areas, both basic and deployed, in which the United States and Soviet Union led in 1976, 1979, and 1986. These figures would be based on differing samples of technology, and they would not be weighted to allow for the fact that some technologies were more important than others. The judgments in individual years about specific technology lags or leads can be questioned. Allowing for all this, the percentage of all technology areas in which the United States led increased from 48 percent in 1976, to 53 percent in 1979, to 59 percent in 1986. The percentage of areas in which the Soviet Union led dropped sharply from 38 percent in 1976, to 20 percent in 1979, to 12 percent in 1986. This corresponds very roughly to the trends in R&D funding, in which the gap between U.S. and Soviet R&D spending opened up around 1970, increased until the middle 1970s, and closed thereafter. This assessment is in rough agreement with another study of technology trends in the 1980s, which argued that out of twenty basic technology areas the

[100]Malcolm R. Currie, *USDR&E Posture Statement for FY 1977*, 3 February 1976, p. II–21.

[101]William Perry, *USDR&E Posture Statement for FY 1980*, 1 February 1979, pp. II–8, II–15, II–17, II–21.

[102]Richard D. DeLauer, *USDR&E Posture Statement for FY 1987*, 18 February 1986, pp. II–11, II–12.

United States improved its standing relative to the Soviet Union in three, lost ground in one, and held its position in sixteen.[103]

This assessment suggests that it may be incorrect to assume that U.S. spending on military research and development led only to a competition in which both sides raced to stay even in a qualitative competition. U.S. spending on technology development relative to that spent by the Soviet Union may have been the critical factor in this regard. Instead of producing a stalemate at comparable levels, increases in U.S. spending on military R&D may have yielded an overall technology lead that not only was stable but also increased over time.

Unfortunately, this assessment is based on evaluations of cost and benefit that are extremely imprecise and subject to error. It is an aggregate judgment and tells us little about which individual research and development decisions have yielded the most value. The judgments expressed in the USDR&E Posture Statement concerned themselves with technological issues, not military utility or effectiveness at the operational level, much less strategic effectiveness in terms of national political goals such as nuclear deterrence. They thus provide little help in reducing the uncertainties surrounding the wisdom of past and future R&D decisions.

Uncertainties about the ultimate efficacy of research and development is matched by uncertainties about the final costs of a research and development program. Andrew Marshall and W. H. Meckling analyzed the problem of predicting the costs of U.S. fighter aircraft, bombers, and missiles developed during the 1950s. They compared the expected and actual costs of weapons in each of these three categories. In their derivations of cost per unit of performance, they adjusted the original cost estimates to correct for inflation in the interval between the initial estimate and procurement and also to allow for modifications of the system. They found that the growth over original estimates averaged out to a factor of 1.8 for fighters, 2.7 to 3.4 for bombers, and 4.1 to 6.4 for missiles. While cost estimates invariably were optimistic, the degree to which they were so varied, and thus a standard correction to initial estimates could not be developed. The conclusion of the authors was that the problems involved in estimating the price of a new weapon that did not have to be modified after work had begun were small compared with the

---

[103]David G. Wegmann, "Net Technical Assessment," NPS-56–89–008 (Monterey, Calif.: Naval Postgraduate School, March 1989, prepared for the Director, Net Assessment, Office of the Secretary of Defense, Washington, D.C.).

[50]

overrun problems created by design changes imposed after the initial cost estimates. They provided no reason to hope that cost overruns could be avoided by keeping the design of new weapons fixed, however, since changes in world politics, enemy capabilities, and available technology that necessitated design changes were more or less inevitable.[104]

Accurate estimates of the cost of new technologies were more difficult the more advanced the technologies were. Further analysis at the Rand Corporation focused on the consequences of using advanced technologies to build new weapons, as opposed to building new weapons that required only the integration of existing technologies. These analyses were undertaken in response to the dramatic growth in the cost of several major U.S. weapons projects in the 1960s. In that period the average cost overrun was 40 percent. What the Rand analysts found, however, was that cost estimates were about equally good for systems initiated in the 1960s as they had been in the 1950s, if one adjusted for the level of technological intricacy of the new weapon. One of the most notorious cases of cost overruns in this period, the F-111 swing-wing fighter-bomber, for example, involved three separate new system designs: Swing wings for efficient flight at high and low speeds, turbofan, as opposed to turbojet, engines for greater fuel efficiency, and radically new onboard electronics systems.[105] In terms of technological difficulty, the Rand analysts ranked the F-111 ahead of weapon systems that seemed more revolutionary to nontechnical observers. At the time of its design the F-111 was judged to have required the invention of more fundamentally new technology than did the first Atlas ICBM or the submarine-launched ballistic missile system. Other systems with extreme cost overruns, such as the Concorde SST, the SR-71 reconnaissance airplane, and the ill-fated first-generation American strategic cruise missiles, the Navajo, Snark, and Bomarc, were even more technologically involved than the F-111.

A strategy for choosing new military technologies, in other words, has to take account of an environment in which it is extremely difficult to make any conclusive analysis of the prospective cost and utility of alternative research and development programs. Furthermore, if the empirical answer to the question of how much was known to the U.S. technological community about foreign military

[104]A. W. Marshall and W. H. Meckling, "The Predictability of the Costs, Time, and Success of Development," in *The Rate and Direction of Inventive Activity*, pp. 461–469.

[105]Coulam, *Illusions of Choice*, p. 39. See also Robert Perry, *System Acquisition Strategies*, R-733-PR/ARPA (Santa Monica, Calif.: Rand Corporation, June 1971), p. 13.

technological innovation is "not much," then we know that American military technological innovation was undertaken in an environment of uncertainty about enemy capabilities, the costs of new technologies, and the benefits of new technologies. The relevant question then becomes one of what strategies the community developed for coping with these uncertainties. This question will be addressed in chapter 8.

CONCLUSION

This chapter began with a set of straightforward and simple questions. I have aimed to show that most of the straightforward and simple answers that we normally take for granted may not be correct. Military organizations do not innovate in peacetime simply in response to defeat or to civilian intervention. Innovation in wartime is not a matter of seeing that existing methods do not work and then correcting them. Technological innovation may or may not be the result of "technology push," "demand pull," or a qualitative arms race with an enemy, and it is certainly not the result of rational calculation of expected costs and benefits. I have offered some alternative approaches that may provide partial answers to these simple questions. Peacetime military innovation may be explainable in terms of how military communities evaluate the future character of war, and how they effect change in the senior officer corps. Wartime innovation is related to the development of new measures of strategic effectiveness, effective intelligence collection, and an organization able to implement the innovation within the relatively short time of the war's duration. Technological innovation is strongly characterized by the need to develop strategies for managing uncertainty. The chapters that follow take up case studies that help flesh out and test these ideas. They are chosen in most cases because they embody the innovations that in large part make the American military what it is today. The U.S. Navy's surface fleet is dominated by the innovation that substituted aircraft carriers for battleships. The United States Marine Corps is dominated by the innovation that substituted amphibious warfare for small war missions. The United States Air Force Strategic Air Command is dominated by the innovations in strategic bombing introduced in World War II and the introduction of guided missiles. Other case studies are chosen because they were classic examples of innovation: the introduction of the tank in World War I, for instance. And still others are studied because they help test some of the ideas developed in this chapter. For example, the role of the Office of the Secretary of Defense in the introduction of helicopter aviation to the United States Army appears to challenge the view that civilian political leaders cannot play a major role in promoting innovation. Similarly, the contributions of scientists to the development of military electronics and guided weapons suggests that civilians with special technical skills can play a significant role in technological innovation. "Bad" innovations, innovations that were put into practice but were clearly mistaken, are not addressed. Despite an extensive and intensive search, no clear-cut cases of bad innovation in the United States military were found. The United States military has made many mistakes, some of which are discussed in this book, but they all appear to have been the result of failures to innovate, rather than inappropriate innovations.

PEACETIME INNOVATION

[2]

The Shape of Wars to Come: Analyzing the Need for Peacetime Innovation

The investigation of peacetime military innovation can begin with a common-sense question. The modern American military is quite noticeably different from the military that emerged from World War I. Innovations must have taken place at some point. How and when did they happen? The following two chapters will argue that peacetime innovations played an important role in shaping today's U.S. Army, Navy, and Marine Corps. Moreover, the pattern of innovation in each of these cases was remarkably similar. Officers within the military developed new ideas about the ways wars would be fought in the future and how they might be won. These new theories of victory were, to a surprising extent, not the product of a close study of potential enemies. Peacetime military innovation in the United States has, in fact, proceeded remarkably independent of intelligence about foreign military powers. Instead, perceptions of change in the structure of the international security environment have been the source of such innovation. The international security environment is composed of those factors not under the control of either the United States military or the government of hostile powers but that constrain or create opportunities for the military. Technological revolutions outside the control of the military, such as the invention of the airplane, or large changes in the international role of the United States, such as the nation's emergence as a Pacific military power after the acquisition of the Philippine Islands, have triggered peacetime military innovation.

It was not unreasonable that these factors, rather than intelligence about potential enemies, were the basis for innovation. As we will see, major changes in the way in which the branches of the

American military fought took many years to accomplish—decades rather than years. Enemy behavior could and did shift much more quickly. An innovation based on the relatively ephemeral matter of enemy behavior might well be found obsolete or unnecessary before it was fully in place. But aviation, for example, has since its invention been a constant of the military environment. Territory acquired by the United States has had to be defended, no matter which party controls the White House.

The process of implementing an innovation has also shown a persistent regularity. Senior military officers who were well respected by traditional military standards have worked to create a new set of operational tasks relevant to the new military capability and a new promotion pathway for young officers to follow as they developed those new skills. Because of the time necessary for young officers to be promoted to senior rank, the practical side of the innovation typically took a generation to accomplish, although senior leadership of the U.S. Army found an ingenious method to shorten this process.

In the sections that follow, the creation of carrier aviation in the U.S. Navy, amphibious warfare in the U.S. Marine Corps, and helicopter warfare in the U.S. Army will be examined. In the period between the world wars, the U.S. Navy transformed itself from a force in which battleships were the dominant weapon to a force capable of executing complex aircraft carrier operations in the first battles of the Pacific war. Contrary to popular conceptions, the American navy was not shocked into abandoning the battleship by the Japanese carrier attack on American battleships at Pearl Harbor. The Japanese carrier fleet was defeated in battle in 1942 and 1943 by American aircraft carriers laid down in the 1920s and 1930s—long before Pearl Harbor.[1] As we shall see, effective doctrine for their use had begun to develop in the same period. Nor was it Billy Mitchell's successful bombing of a captured German cruiser, the *Ostfriesland,* on a test range in 1921 that pushed the navy into purchasing carriers. The service had submitted budget requests to Congress for two aircraft carriers more than a year before the sinking of the *Ostfriesland,*[2] and the General Board, the navy's most senior advisory council, had spent much of 1919 in secret hearings debating the future of navy aviation, concluding the "fleet aviation must be developed to the fullest. . . . Fleet engagements of the future will be

---

[1]Clark Reynolds, *The Fast Carriers: The Forging of an Air Navy* (Huntington, N.Y.: Robert Krieger, 1978), pp. 18–19, 38, 83.
[2]Charles M. Melhorn, *Two Block Fox: The Rise of the Aircraft Carrier 1911–1929* (Annapolis: Naval Institute Press, 1974), pp. 64–65.

probably preceded by air engagements. . . . Development of all types of aircraft . . . and fleet aviation are the most important work for the immediate future."[3]

In the same period, the U.S. Marine Corps went from a force that had fought small wars in Asia and the Caribbean and as infantry in Europe in World War I to the first force capable of amphibious assault on heavily fortified island targets. The creation of amphibious warfare by the Marine Corps was a major innovation recognized as such during and after World War II. The military innovator and historian J. F. C. Fuller went so far as to write that Marine Corps operations in the Pacific were "in all probability . . . the most far reaching tactical innovation of the war."[4]

The addition of the airmobility afforded by the helicopter to the combat capabilities of the army was a less sweeping innovation than those of the navy and the Marine Corps, which involved not just the addition of a new force but a radical readjustment of the role of older combat branches. However, the army innovation was effected much more rapidly than those in its sister services. In the 1950s, army aviation attached to ground units was transformed from a small, noncombat arm used mostly for transporting senior officers to a helicopter-mobile assault force capable of facing heavily armed opponents and Vietnamese communist regular and irregular troops. The first experimental Sikorsky military helicopters were not delivered to the army until 1942, and they were first used for medical missions by the Coast Guard in 1944. The first helicopters delivered to combat units appear to have been thirteen produced by the Bell Company and delivered to the 82nd Airborne Division for test and evaluation in 1946.[5] The Army used helicopters on a very limited basis for medical evacuation in the Korean War[6], but their role in combat was not developed for almost a decade. General Hamilton Howze, director of army aviation from 1955 to 1957 and president of the Howze Board that formulated Army air-mobility requirements in

---

[3]Cited in Thomas C. Hone, "Navy Air Leadership: Rear Admiral William A. Moffett as Chief of the Bureau of Aeronautics," in *USAF Warrior Studies—Air Leadership: Proceedings of a Conference at Bolling Air Force Base April 13–14* (Washington D.C.: GPO, 1984), pp. 89–90.

[4]Quoted in Jeter Isely and Philip Crowl, *The U.S. Marines and Amphibious War: Its Theory and Its Practice in the Pacific* (Princeton: Princeton University Press, 1951), pp. 6, 8.

[5]Igor Sikorsky, *The Story of the Winged-S* (New York: Dodd, Mead, 1967), pp. 219–20; James Gavin, *War and Peace in the Space Age* (New York: Harper, 1958), p. 109.

[6]John J. Tolson, *Vietnam Studies: Airmobility 1961–1971* (Washington, D.C.: GPO, 1973), p. 8.

1962, stated that he did not think, in 1953 and 1954, that "the Army . . . had recognized at all the potential of the helicopter. There was very little knowledge in the Army that the Marines during one very small operation in Korea had lifted by helicopter a small party of Marines to the top of a mountain, an unoccupied mountain, and had found this a very useful thing to do."[7] Yet the first Calvary Division (Airmobile), using helicopters for the transport of troops and artillery in combat operations, was engaged in combat in South Vietnam by 1965. The army thus went from nothing to a new combat force in little more than a decade.

This is a study of success, and the main message, which is more important than the analysis of how the innovations were accomplished, is simply that farsighted peacetime military innovation has been possible in the modern American military, even during the 1920s and 1930s when military budgets were tight and popular attitudes toward the military were far from friendly, and even in the 1950s, when the military bureaucracy had swollen in size far beyond prewar levels. But equally important is the concluding section, which is a study of failures, of efforts to promote innovation which were initiated but which failed to generate useful capabilities. The patterns of action found in the cases of successful innovation are noticeably absent in the cases of failed innovation.

INTELLIGENCE AND MILITARY INNOVATION

What creates the perception of a need for innovation? A reasonable hypothesis would be that intelligence about the military plans and capabilities of potential enemies drives military planners to develop countermeasures in the form of new weapons and concepts of operations. It might be expected, for example, that intelligence about Japanese naval warfare plans and capabilities in the Pacific drove the United States Navy to develop the concept of independent aircraft-carrier strike forces and the Marine Corps to develop the capability for amphibious warfare. One account of the origins of helicopter aviation in the U.S. military argues that it was the sight of masses of helicopters carrying infantry at the 1956 Moscow air show

[7]Interview with General Hamilton H. Howze, USA (ret.), by Colonel Glenn A. Smith and Lt. Colonel August Cianciolo, *The History of Army Aviation, Senior Officer Debriefing Program* (Carlisle Barracks, Pa.: U.S. Military History Institute), p. 8; available at the U.S. Army Center for Military History, (CMH), Washington, D.C. (interviews in this series hereinafter cited as *Debriefing, CMH*).

that galvanized the Army into developing analogous capabilities.[8] The historical record, however, seems to suggest that U.S. intelligence about the Japanese was too imprecise and too unstable to serve as an intellectual basis for military innovations that might take decades to accomplish. Intelligence about the Soviet Union also seems not to have been a major factor in the development of helicopter warfare in the army.

At the time the U.S. Navy began to consider aviation for military missions and the future of Marine Corps operations in the Pacific, U.S. naval intelligence was not able to provide data or analysis that might help determine future strategic requirements. The available evidence suggests that up until 1927, U.S. intelligence about the Imperial Japanese Navy was based entirely on the reporting of attachés who sifted through Japanese newspapers, government documents, and interviews with Japanese officials. Their haul was usually poor, and the director of naval intelligence, Luke McNamee, complained to one of his attachés in Tokyo in 1923 that the Office of Naval Intelligence was unable, for example, to determine the types, numbers, and characteristics of Japanese submarines and that it lacked detailed information about those Japanese warships regulated by treaty, the Japanese shipbuilding program, and the technical characteristics of the largest Japanese battleships.[9]

Matters improved somewhat after 1927, when a copy of the Imperial Japanese Secret Operations Code–1918 came into the possession of the Office of Naval Intelligence, was translated, and was used to break the cyphers protecting Japanese radio transmissions.[10] Technical intelligence about the capabilities of Japanese warships showed particular improvement.[11] Until 1930, however, the bulk of U.S. naval intelligence reporting on the Japanese military relied on open sources. The reporting on the 1927 Japanese Grand Fleet maneuvers,

---

[8]John R. Gavin, *Air Assault: The Development of Airmobile Warfare* (New York: Hawthorne, 1969), p. 274.

[9]Jeffrey M. Dorwart, *Conflict of Duty: The U.S. Navy's Intelligence Dilemma 1919–1945* (Annapolis: Naval Institute Press, 1983), pp. 26, 28.

[10]L. F. Safford, "The Undeclared War: History of R.I. [Radio Intelligence]," SRH-035, 15 November 1943, Layton Papers, Historical Collection, Naval War College, Newport, R.I., pp. 3–4.

[11]See, for example, in "Various Reports on Japanese Grand Fleet Maneuvers 1927–1929," SRH-320, the report from the *USS Pittsburgh* to Director Naval Intelligence, 28 November 1927, Record Group 457, National Archives, Washington, D.C. The speed of Japanese warships in trials was reported by radio and was intercepted by the U.S. Navy, for example, in 1936. See Edwin P. Layton, with Roger Pineau and John Costello, *"And I Was There": Pearl Harbor and Midway—Breaking the Secrets* (New York: William Morrow, 1985), pp. 57–58.

for example, was based on Imperial Japanese Navy press releases and ship movement reports. This information did not add up to a picture of a Japanese fleet armed with aircraft carriers that could be countered only with American aircraft carriers. On the contrary, the dominant impression was of a Japanese navy intent on wearing down the U.S. Navy by drawing it into operations within range of Japanese ground-based aircraft based in Japan or in the Philippines after those islands had been captured by the Japanese. The logical response would be for the American fleet to stay out of range of Japanese ground-based aircraft.[12] Reports on lesser Japanese exercises seemed to demonstrate only that the Japanese were concerned with the possibility of an attack by carrier-launched bomber aircraft against Japan and with vulnerability of the ground-based aircraft that would be used to defend Japan.[13] By 1930, in other words, there were intelligence reports indicating that the Japanese were worried about U.S. innovations in carrier aviation, but not that the United States needed to react to Japanese developments.

The American ability to intercept and decrypt Japanese radio communications had improved enough by the time of 1930 Grand Fleet maneuvers for the United States to detect the mobilization of the Japanese fleet, track its movements and the state of its fuel supplies, and determine once again that the Japanese navy was practicing for a war in which American aircraft carriers approached close enough to the Japanese home islands to attack Tokyo. Unfortunately, even this limited intelligence was not reported back to the Office of Naval Intelligence for three years, because the U.S. admiral in charge of the intelligence-gathering mission "felt he could not trust the Acting Director [of Naval Intelligence] and put a taboo on the whole office," just to play safe.[14] When the data was finally analyzed, it was determined that the 1930 exercise was a cover for the operations conducted in 1931 to protect Japan's seaward flank as it invaded Manchuria.[15]

The Japanese Grand Fleet maneuvers of 1936 might have yielded more intelligence on the strategic challenge posed by the Japanese to

[12]SRH-320, attaché report 332, 23 November 1927 on Japanese Grand Fleet Maneuvers, 11–24 October 1927.

[13]SRH-320, attaché report 167, 4 October 1929, "Japanese Air Maneuver"; report 192, 4 December 1929, no title.

[14]Safford, "Undeclared War," pp. 6, 8, 10.

[15]Lawrence Safford, "A Brief History of Communications Intelligence in the United States," SRH-149, reprinted in Ronald Spector, *Listening to the Enemy: Key Documents on the Role of Communications Intelligence in the War with Japan* (Wilmington, Del.: Scholarly Resources, 1988), p. 6.

the U.S. Navy. Two carrier-based aircraft squadrons were involved in the exercise's attacking force, without any detectable participation of battleships or cruisers. In retrospect, this suggests that the Japanese may have been developing an independent strike role for their aircraft carriers. But at the time interpretation was hindered by a shortage of data. The bulk of the maneuvers were held close to Japan in areas where the U.S. had little ability to intercept radio communications.[16]

The best U.S. naval intelligence suggested that the Japanese navy had developed carrier aviation but was particularly interested in the defensive uses of air power, particularly ground-based air power, against American carrier-based craft trying to bomb Japan. The Japanese acted as though they believed the war would be fought close to Japan and that Japanese carriers would not be used as platforms for long-range strikes against American ships or land bases. There was a long-range role for Japanese aircraft carriers, but, as I will show, that followed, rather than preceded, American innovations in this area. In short, American intelligence provided no indication that there was a Japanese carrier aviation threat, or any other specific Japanese military capability, that had to be matched or countered by an American carrier force. As far as concepts of amphibious warfare were concerned, Japanese military exercises seemed overall to be geared toward operations close to home. It was clear that the Japanese meant to seize the Philippines Island however, and this did spur U.S. interest in amphibious warfare concepts. But there do not appear to have been specific Japanese military plans or capabilities that highlighted a need for new Marine Corps capabilities.

Nor can the genesis of army helicopter warfare be seen as a response to intelligence about enemy plans and capabilities. The Howze Board report, which has been referred to as the formal source of the army's creation of helicopter assault forces, does have a lengthy intelligence annex, but it is simply a general survey of all Soviet and Chinese military capabilities and does not focus on Soviet helicopter aviation or any new Soviet capability that created a special need for an American helicopter assault force. No special strategic problem requiring the use of helicopters is mentioned, and the board offered only the unexceptionable judgment that the likelihood of a limited war resulting from Soviet aggression was high "where

---

[16]"U.S. Navy Reports on Japanese Grand Fleet Maneuvers," SRH-318, Commander in Chief Asiatic Fleet to Chief of Naval Operations, 13 March 1937, Record Group 457, National Archives, Washington, D.C., pp. 026, 029.

[63]

opportunities are great and the risks relatively low" and gave a very general picture of the current situation in various trouble spots around the world. There is only a one-paragraph description of Soviet helicopters.[17] No attempt was made to link the material in the intelligence annex to the recommendations for expanded helicopter procurement and use, and the final report itself has no substantive discussion of intelligence issues at all, noting merely that "the headquarters reviewing this report are entirely current on the latest intelligence of the Sino-Soviet bloc; the information is therefore omitted here."[18] As to the relevance of the 1956 Moscow air show, it will be shown that most of the key decisions to develop helicopter aviation were made before 1956.

THE ENVIRONMENT, SIMULATION, AND INNOVATION

If it is difficult to uncover links between specific enemy plans and capabilities and military innovation in peacetime, is it possible to determine what led officers to perceive the need for major changes in their organization? Intelligence about potential enemies was uncertain, and the potential enemy's capabilities and plans could change rapidly in response to changes in policy. Thus such intelligence might or might not provide a reliable indication of the long-term need for new military capabilities. Study of the changes in the economic, political, or technological realms that were beyond the control of governments and that constituted the environment with which military organizations had to contend could provide a more stable basis for deciding whether military innovation was necessary and what its character might be.[19]

The slow march of the Marine Corps toward amphibious warfare suggests that changes in the security environment did trigger the innovation. The acquisition of the Philippine Islands in the Spanish-American War appears to have been the critical event that led to a force for seizing and defending unoccupied positions for use as advanced naval bases. This in turn led to the perceived need to be able to seize positions occupied by the enemy for use as advanced naval

---

[17]Annex D, "Intelligence," to *Final Report U.S. Army Tactical Mobility Requirements Board,* Fort Bragg, N.C. 20 August 1962 (hereinafter *Howze Board*), CMH, pp. 2–4, 9–12, 15–25; discussion of Soviet high-speed helicopters on p. 23.

[18]*Howze Board,* p. 1.

[19]To the best of my knowledge, the concept of the security environment was originated by Andrew W. Marshall.

bases. The navy made repeated efforts to obtain from Congress funding for advanced-base material. After these efforts failed, Admiral Dewey, in 1905 the president of the navy's General Board, carefully outlined the strategic rationale for the new force: "[A]ll the war plans made by the General Board demand an advanced base of operations, the precise location, defenses, and time of occupation of which depend on the circumstances of the particular campaign. In some combined operations the plans prescribe that the advance base shall be seized and held by the Navy and Marines. . . . The rendezvous of the Japanese fleet during the blockade of Port Arthur, which was defended by mines, booms, etc. . . . is a good example of what is meant by an advanced base."[20]

Dewey elaborated on this theme the following year. The acquisition of the Philippines created the potential for naval wars between the United States and European and Pacific Ocean powers. In the event of such a war, at least one of the navies, and perhaps all, would have to operate far from home bases. Advanced operating bases would be needed so that the fleets would not have to return to home ports for repair and resupply. The Japanese had acquired and utilized just such an advanced base at Masampho, Korea, during the blockade of Port Arthur. It might or might not be necessary for the U.S. to have an advanced base in the Atlantic, on the Brazilian coast, for example, if it went to war against a European power. It would certainly need one in the Pacific, given the distances from the United States to the Asian coast. Hence the need for a force that could establish these advanced bases.[21] Other navy and Marine Corps officers developed this concept, explaining that while defended islands would not be seized, the bases so established would have to be able to defend themselves against enemy naval raids without help from the U.S. Navy. These bases would be to increase the flexibility of the navy. If ships had to be tied to these new bases to defend them, the whole purpose of the endeavor would be lost.[22]

[20]Admiral Dewey to the secretary of the navy, 3 February 1905, Reports of the General Board, File 408, microfilm collection, Historical Collection, Naval War College.

[21]Admiral Dewey to the secretary of the navy, 3 August 1906, Reports of the General Board, File 408, microfilm collection, Historical Collection, Naval War College.

[22]USMC Major Dion Williams, "Report on Men, Material, and Drills Required for Establishing Naval Advanced Base," 2 November 1909, report by the Office of Naval Intelligence, Reports of the General Board, File 408; Admiral Dewey to the secretary of the navy, "Detail Plans for Advanced Base Operations Next Winter," 21 July 1913, Reports of the General Board, File 408, microfilm collection, Historical Collection, Naval War College.

Far from embracing this new mission as a way of expanding the resources they received, the marines and the navy went slow. The marines were busy fighting a small war in the Caribbean, and there was fear among conservative corps officers that creating a permanent force dedicated to this mission would result in the Marine Corps being taken away from the navy by Congress. Exercises were held that demonstrated that the marines could seize an unoccupied island and defend it against a force trying to take it back. By World War I, however, no steps had been taken toward establishing a force that specialized in this capability or toward establishing one that could seize a defended island.[23]

The transition from a force that could defend to one that might seize such bases from the enemy appears to have been the work of one man, Major Earl H. Ellis, USMC. An officer who had worked closely in World War I with the future commandant of the Marine Corps, John A. Lejeune, Ellis was the author of a 1921 study that observed concerning the Pacific: "All ports suitable for use as advanced bases by the United States fleet will be denied in some strength. This will necessitate the execution of opposed landing operations." He noted the obvious difficulties that an attack against a defended beach would involve but stopped short of discussing how those difficulties could be overcome.[24]

Ellis then took the next intellectual step, of translating the perceived need for a new capability into concrete, new military tasks. He wrote Operations Plan 712D, which was endorsed by Commandant Lejeune as part of the war-plan portfolio approved in 1921. It is concerned for the most part with the older mission of defending advanced bases. But in eighteen pages it sketched out an entirely new form of warfare and identified the new tasks that its conduct entailed. Having reviewed earlier landings, he concluded that with machine guns and artillery "on the side of the defenders, and only limited places available to the attackers, it would be a desperate undertaking and only carried out at great sacrifice." Therefore, an

---

[23]See Graham Cosmas and Jack Shulimson, "The Culebra Maneuver and the Formation of the U.S. Marine Corps' Advanced Base Force 1913–1919," in Merrill Bartlett, ed., *Assault from the Sea: Essays on the History of Amphibious Warfare* (Annapolis: Naval Institute Press, 1983), pp. 121–32; Aide for Inspections Captain William Fullam to the Secretary of the Navy, "A Flotilla to be organized in connection with an Advanced Base outfit," 28 June 1913, with comment by Dewey dated 21 July 1913, General Board, microfilm collection, Historical Collection, Naval War College.
[24]Earl Hancock Ellis, "Naval Bases: location, resources, denial of advanced bases," typescript, Naval War College Library, 1921.

attacking force would have to have extremely high morale, self-confidence, and familiarity with traversing daunting terrain. The assault would have to be executed with rapidity and continuity to keep the defender off balance, and aerial strafing and naval gunfire would have to be applied to keep the defenders suppressed while the assaulting troops were wading ashore. Traditional concepts of ground warfare, such as the preparation of artillery bases on shore before moving the assault inland and the gradual wearing down of fortified positions, all had to be rejected in favor of the rapid execution of preplanned, high-speed assaults. Traditional concepts of amphibious warfare that avoided the firepower of the defense by landing at night or on undefended portions of the island were also rejected, since many Pacific islands were too small to have undefended sectors and a nighttime assault of thousands of men was in practice unmanageable.[25] Ellis had successfully identified the essence of the concepts that would be used by the marines to assault islands in the Pacific in World War II. In his plan he even correctly estimated the size of the force that was actually used to seize one island. He did so by breaking radically with the conventional wisdom for both ground force and amphibious operations. He was an authentic military genius.

What does this example of the intellectual process leading to the conceptual basis of a military innovation tell us? It is reasonably clear that the perceived need for a new capability had its roots in the emergence of the U.S. as a global naval power, particularly in the Pacific. The political need to defend newly acquired national interests meant that U.S. forces would have to operate in a new security environment. Instead of defending the continental United States alone, naval forces would not need to be able to operate at long distances from the United States for extended periods of time. An examination of the new environment then suggested that the bases we would need would have to be fought for, not simply occupied. The individual brilliance of Earl Ellis then translated that perceived requirement into new military tasks. It is not possible to explain individual genius, but the structural factors that led the U.S. Navy and Marine Corps as organizations to an appreciation of the need for amphibious warfare are reasonably clear.

[25]"Advanced Base Operations in Micronesia," signed by John Lejeune, 23 July 1921, photocopy from the Historical Amphibious File, Breckinridge Library, Marine Corps Schools, Quantico, Va., pp. 18, 22, 29.

Did perceived shifts in the security environment spark the development of carrier aviation? How was the need for a new capability translated into new military tasks? To answer these questions we must review the initial impact of aviation on naval operations. The invention of aircraft was followed rapidly by its use in World War I by the navies of the United States and Great Britain. What was immediately obvious was that aircraft increased the range at which the enemy fleet could be observed. Aircraft helped navies better perform an old mission, scouting for the battleships. Aircraft could also observe and direct the gunfire of battleships against enemy targets, helping them to be more accurate. Aircraft carriers, aircraft launched from battleships and cruisers, and seaplanes could all be used to help the battleship fight its battles better. This use of aircraft was not innovative, did not challenge existing roles and missions within the Royal Navy or the U.S. Navy, and was quickly implemented during World War I. The real innovation revolved around the question of whether aircraft carriers were to remain "the eyes of the battleship" or would constitute an independent strike force that would replace the battleship as the dominant naval weapon.[26]

Although visionaries like Billy Mitchell were willing to argue that aircraft had already overtaken the battleship as the dominant naval weapon, the facts were that in the early 1920s the capabilities of aircraft were not sufficiently developed for this to be true. The *Ostfriesland* had been sunk by aircraft, but it had been tethered, unarmed, and unprotected by escorts or aircraft of its own. There was not yet any combat experience by which to judge the feasibility or utility of aircraft carriers conducting independent missions. There were not even any realistic fleet exercises that could be used to test the validity of a new role for aircraft carriers. Fleet exercises in 1923 had to use battleships as nominal aircraft carriers and single aircraft

---

[26]For example, while chief of the Bureau of Aeronautics in 1928, Admiral William Moffett forwarded to the secretary of the navy the views of the commander in chief, U.S. Fleet, endorsing the use of naval aviation as an adjunct to battleships: "In the absence of an aircraft carrier for the battle line on which *gunnery observation planes* may land . . . the Fleet is training for war under a handicap which must be realized." The commander in chief, Battle Fleet chimed in: "Experimental practice was fired by a division of three ships. . . . The opening range, bearing of targets and all spots were obtained from *observation planes* with surprising accuracy. . . . The results of this practice demonstrate the practicability of engaging the enemy beyond the horizon, at long range under low visibility conditions, or behind a smoke screen or high land"; emphasis added. See "The Influence of Developments in Naval Aviation on the Development of the Art and Material of Naval Warfare," 17 December 1928, endorsing Moffett's memo of the same title to the secretary of the navy, 31 July 1928, File 449, Records of the General Board, National Archives, Washington, D.C.

had to represent squadrons, because real carriers and airplanes were not available.[27]

Simulations of naval warfare, however, could be used to extrapolate technological trends in the strategic environment. No navy in the world could put two hundred aircraft out to sea on aircraft carriers in the 1920s. But what if they could? Would an independent offensive role for carriers be possible under those conditions? The key to the intellectual breakthrough appears to have been such simulation. Many of the most useful carrier warfare exercises were conducted against a simulated Red, or British, fleet in the Atlantic as it moved toward the coast of North America, as well as against the Orange, or Japanese fleet. What was important was the ability to simulate in war games the impact of potential roles of aviation.

For example, one war game played at the U.S. Naval War College in Newport, Rhode Island, in the fall of 1923 employed a Blue, or American, fleet with five aircraft carriers, more carriers than any country in the world then possessed, against a Red fleet with four aircraft carriers. Although these carriers were employed along with battleships, they explored the implications of radical increases in the airpower available to navies. Blue aircraft carriers in this simulation launched two hundred aircraft in one strike at the Red fleet, employing torpedos and bombs, and succeeded in crippling Red's aircraft carriers and one battleship. Massive air-to-air battles involving over one hundred friendly and enemy aircraft were fought in the skies over the Blue fleet. Most important, concepts essential to the conduct of carrier warfare were worked out. The necessity of massing aircraft for strikes was highlighted. Rather than assigning aircraft to each battleship to act as its eyes, they were launched and kept in the air until large numbers could be assembled for an independent strike. The need for a coherent air-defense plan to coordinate the use of defensive aircraft was emphasized, and the commander of the Red fleet was faulted for failing to come up with such a plan. Control of the air was established as the first goal of air operations.[28] In a game played in 1925, the Blue fleet commander was severely criticized for his inability to formulate a coherent plan

[27]Archibald D. Turnbull and Clifford Lord, *History of United States Naval Aviation* (New Haven: Yale University Press, 1949; rpt., New York: Arno Press, 1972), p. 214.
[28]Captain Harris Laning, Tactical Problem II (Tac 10 Mod 9), "Battle of Sable Island," History and Tactical Critique, October–November 1923, Class of 1924, Record Group 12, Historical Collection, Naval War College, pp. 5, 28–30, 38–40, 44.

for the use of this air power.[29] Additional war games over the next six years suggested that carrier air strikes could also cripple Japanese battleships in the central Pacific.[30] By 1935, scoring rules for war games at the Naval War College assessed the impact of aviation in naval warfare as 25 percent of the total, more than maneuver and gunfire put together.[31] Though these games continued to feature engagements between battleship fleets accompanied by aircraft carriers, they clearly suggested the growing importance of carrier aviation and helped push the navy toward an essential intellectual reformulation of the character of naval warfare.

Rear Admiral William Moffett was named as the first chief of the navy's Bureau of Aeronautics in 1921. He drew on the results of these simulations in developing a clear picture of an independent role for carriers.[32] His initial concern was primarily to obtain a building program that would provide an adequate number of carriers, and he endorsed the use of carriers as scouts for battleships in order to build a broad constituency within the navy for their construction. In those early years, however, he was also careful to campaign for many different and specialized types of carrier aircraft, so that the full potential of each—fighters and bombers as well as scout planes—could be determined and so that carriers would not be forced into serving only as scouts for battleships because that was the only mission their aircraft could fulfill.[33] By 1931 he was in a position to present his reformulation of the character of naval operations in a memo to the secretary of the navy. Carriers should no longer be considered as adjuncts to battleships, but as replacements for battleships: "The function of a large carrier should be the same as that of a battleship, viz., to deal destructive blows to enemy vessels. Its offensive value is too great to permit it to be ordinarily devoted to scouting." The best evidence for this assertion at that time was the result of the simulated engagements performed in war

[29]Tactical Problem II (Tac 98), Blue-Red Tactical Exercise, 14 January 1925, Class of 1925, Record Group 12, Historical Collection, Naval War College, pp. 17–21.
[30]Tactical Problem II-1932-SR, "History of Maneuver," October 1931, Record Group 12, Historical Collection, Naval War College.
[31]Michael Vlahos, *The Blue Sword: The Naval War College and the American Mission, 1919–1941* (Newport, R.I.: Naval War College Press, 1980), pp. 136–37.
[32]Melhorn, *Two-Block Fox*, p. 89; 1927 statements by the president of the Naval War College quoted in Moffett, "Influences of Developments in Naval Aviation."
[33]William Moffett to the Chief of Naval Operations, "Aircraft Building Program—Comments on Memorandum Submitted by the Head of the War Plans Division," 17 December 1928, File No. 449, Records of the General Board, National Archives, Washington, D.C.

games. War games also provided the basis for his bold redefinition of carriers themselves.

> An aircraft carrier is a flying field at sea . . . merely a means of launching and landing bombing, torpedo, scouting, or fighting airplanes. As a flying field, it should not place itself willingly where it can be attacked, especially by gunfire. . . . Its primary mission is to make possible the operation of airplanes at sea. . . . Its gun defenses should be provided from sources outside itself, by cruisers, battleships, and destroyers. Nothing should interfere with its primary mission—furnishing a means of operating aircraft as rapidly as possible. It is a means of furnishing great offensive power if nothing is allowed to interfere with its primary mission.[34]

This reformulation of the role of the carrier proceeded from the perception of trends in aviation technology that were highlighted in simulations. The reformulation led directly to a new set of military tasks. If an aircraft carrier was not primarily a part of the battle fleet, to be maneuvered as part of that fleet, but a floating airfield, it would have to maneuver to take advantage of the wind in order to launch and recover airplanes most easily. The fleet would have to follow the carrier, not the other way around. Carrying and launching as many aircraft as possible became the new critical task of aircraft carriers. It was not by accident, therefore, that Admiral Joseph Mason Reeves, a battleship admiral who was an ally of Moffett, ordered the *Langley*, the first American aircraft carrier, to take on board forty-two aircraft, more than three times her normal complement, without the knowledge and against the wishes of her captain, prior to the 1928 fleet exercises off Hawaii.[35] At that time, the Royal Navy was not able to operate more than a dozen aircraft from a carrier, and British visitors refused to believe that the *Langley* operated with even twenty-four aircraft.[36]

A perceived change in the character of the security environment and simulated combat also provided the intellectual basis for the development of helicopter aviation in the U.S. Army. In this case, the

---

[34]Moffett to the president of the General Board, "Design of Future Aircraft Carriers," 12 November 1931, Serial Number 1554, File No. 420, microfilm collection, Historical Collection, Naval War College.

[35]Eugene E. Wilson, "The Gift of Foresight: The Reminiscences of Commander Eugene E. Wilson," Columbia University Naval History Project, Oral History Research Office, 1962, typescript, pp. 354–371. Wilson served on the *Langley* before going on to a career in the aviation industry.

[36]Hone, "Navy Air Leadership," p. 102.

change was due to technology rather than increased international political commitments. General James Gavin began writing in 1947 about the army's need for strategic airmobility, the ability to transport units by airplane between strategic theaters of operations, and tactical airmobility on the battlefield using aircraft that could take off and land in short distances.[37] It was the advent of nuclear weapons and ballistic missile delivery systems that motivated Gavin. The Soviet Union did not yet have nuclear weapons, but Gavin was concerned with a general trend in the security environment. Firepower was becoming available to all modern nations in ever increasing amounts, most dramatically in the form of tactical nuclear weapons. This trend toward greater firepower was not being matched by an increase in the mobility of military units. While Gavin was the commanding officer of the American VII Corps in West Germany, one of Gavin's subordinate officers noted that his chief was worried by the vulnerability this created: "If a unit was to concentrate sufficient force so that it would be able to perform decisively on the battlefield, it would present a very lucrative target to enemy nuclear weapons, and . . . therefore, the tactics of operations required a continuous massing and decentralization of the forces. . . . [While] you could do it [mass and decentralize] on a map exercise . . . to do it practically on the ground was not feasible."[38]

Greater firepower on the battlefield, including chemical as well as nuclear weapons, created a vulnerability. Simulations, in the form of map exercises, indicated that increased mobility would be necessary to reduce this vulnerability. But simulations did not indicate how to solve the problem in practice. In the real world, in which movement is blocked by waterways, forests, and mud, the rapid concentration and dispersion that was simulated in map exercises was impossible, at least with the ground transportation then available or imaginable. After leaving Germany in 1954, Gavin became the head of the Army General Staff's Section G-3 for Plans and Operations. The officers working for Gavin in that office had been exposed in Korea to the experimental use of helicopters by the Marine Corps to transport troops into combat on the battlefield, and they realized that the helicopter was the practical answer to the conceptual requirement identified by Gavin's map exercises. The need for greater mobility, created by perceived changes in the security environment and rein-

---

[37]Gavin, *War and Peace in the Space Age*, pp. 107–9.

[38]Interview with Brigadier General E. L. Powell, USA (ret.), by Colonels Bryce R. Kramer and Ralph J. Powell, 18 March 1978, *Debriefing*, CMH, pp. 10–14.

forced by simulations of new concepts of operations, triggered the search, which finally fixed on helicopter aviation. Simulations were an important link between perceptions of a new environmental problem and possible practical solutions. New solutions could not yet be tried out in the field, because helicopter technology had not yet progressed far enough.

The concern with greater mobility led directly to Gavin's choice of Hamilton Howze to be the chief of Army Aviation in charge of developing helicopters for combat use. This appointment created some puzzlement and disappointment among army aviators, because Howze was a tank commander with no aviation background. Howze recalls that he "was told by Jimmy Gavin that he had selected me purely by reason of the fact that I had tremendous interest in mobility as being the real key to battlefield success."[39]

The trend toward greater firepower created the need for greater mobility, but so did the irregular warfare that was anticipated in the poor rural areas of the world, and this also spurred the development of helicopter aviation. The Howze Board report identified the trend toward high accuracy missiles, chemical and biological weapons, and improved nonnuclear weapons as one set of factors increasing the need for mobility. But ground combat against guerrillas in Asia or Africa also had to be considered. "In the circumstances [there] . . . surface mobility is very often that of a marching soldier . . . and the guerrilla . . . must be credited with a foot mobility and an elusiveness greatly exceeding that of his regular soldier opponent. . . . We must redress the imbalance, and since the terrain and the nature of the fighting give little hope of doing it by ground vehicle, we must . . . take to the air."[40] Since helicopters were expected to be vulnerable to sophisticated anti-aircraft weapons for some time to come, helicopters might very well prove useful against guerrillas before they could be used on the European battlefield.[41]

Once the general argument for helicopter mobility had been made, a large question remained. What would warfare incorporating helicopters look like? In a spate of individual brilliance Major Ellis had worked out the essential aspects of amphibious assault warfare. In this case, however, and as with carrier aviation, the intellectual task of going beyond a general understanding of the potential utility of an innovation to a description of how that

---

[39]Interview with Howze, *Debriefing*, CMH, p. 1.
[40]*Howze Board*, pp. 11–12.
[41]Annex K, "Tactical Concepts and Requirements," *Howze Board*, pp. 1–3, 22–23.

innovation would be implemented was performed collectively by the service. As in the case of carrier aviation, it was not possible to stage exercises with men and equipment to work out the shape of the new way of war, because the necessary machinery was not yet available. Helicopters in the early 1950s were unreliable, unable to carry large cargoes, and available only in small numbers. Once again, simulation in the form of war games proved useful. If real helicopters were not available for use in battlefield exercises, notional helicopters would be used. Howze, while director of Army Aviation,[42] took tactical map problems then in use at the Army Command and General Staff College at Fort Leavenworth and modified the forces employed in them to include helicopter air assault units that could transport troops to rapidly exploit tactical opportunities. In his words, "It was an extremely effective device simply because it wasn't something that I had dreamed up from the start. I said 'This is a problem now taught at the Infantry School. Now if you give him ten [helicopters] and a couple of armed helos, this is what he can do.'"[43]

This simple war game exercise of elaborating the operational implications of helicopter mobility for combat units was eventually supplemented by field exercises at Fort Rucker, the home of the Army Aviation School, using real helicopters to test experimental tactics that have since become familiar, such as "nap of the earth" flying and "pop-up" tactics to avoid enemy radar and basic reconnaissance techniques.[44] The advocates of the helicopter explicitly warned, however, that these exercises were only simulations. They could merely suggest what future wars involving helicopters might look like and how the new weapon might best be used. They could not predict what wars would look like nor could they "prove" that the helicopter would work. A radically new weapon could not be dropped into war games that had been designed to simulate traditional forms of warfare without introducing large uncertainties into the simulation. As the Howze Board noted of these exercises, "Much of the [evaluation] work had to be done on a pretty subjective

---

[42]Howze's office and title were upgraded during his tenure.

[43]Interview with Howze, *Debriefing, CHM*, pp. 19–20. See also the description of the work of Colonel J. D. Vanderpool at the Combat Development Center at Fort Rucker in 1956 and the work of William Howell at Fort Benning in developing operational tasks for armed helicopters described in Frederic A. Bergerson, *The Army Gets an Air Force: Tactics of Insurgent Bureaucratic Politics* (Baltimore: Johns Hopkins University Press, 1980), pp. 72–76.

[44]Interview with E. L. Powell, *Debriefing, CMH*, p. 33.

basis."[45] Simulating new forms of warfare will always be full of uncertainties, because there is no reality against which to test the simulation. Yet there may be no better way to think through innovative practices in peacetime.

Conclusion

Each of the cases examined suggests that military planners were driven to consider the need for innovation by broad structural changes in the security environment in which their organizations would have to fight for the foreseeable future, not by specific capabilities or intentions of potential adversaries. Working out the specific character of the innovation was usually a matter of simulating the character of future warfare. Field exercises could not be conducted with weapons that did not yet exist or existed only in nascent form. Simulations were thus useful in thinking through the shape of potential innovations. This method carried risks, since simulations are not predictions, though they might be taken for such, and so cannot foretell the efficacy of an innovation on the battlefield. They can only suggest which approaches needed further investigation and development.

Once the need for an innovation was perceived, and the character of the new capabilities conceptualized, the organization had to be made to take action. The behavior of the organization, as it prepared for and conducted war, had to change if the innovation was to be actualized. The intellectual innovation had to be accompanied by a political process that changed the way the officers and men of the military lived their professional lives. How that happened is the subject of the next chapter.

[45]*Howze Board*, p. 6. See also Annex M, "War Games," to the *Final Report* for a fuller discussion of the artificialities of war games for innovative weapons and military concepts.

[3]

Making Things Happen: The Politics of Peacetime Innovation

Perceptions of structural changes in the security environment and simulations of hypothetical wars involving new military capabilities appear to have been at the heart of the intellectual development of American military. Identifying new strategic requirements and translating them into new tasks is, however, only half the battle toward innovation in a bureaucratic organization. Analysis of the military organizations discussed in the previous chapters suggests that innovations occurred when senior military officers were convinced that structural changes in the security environment had created the need. These senior officers, who had established themselves by satisfying the traditional criteria for performance had the necessary power to champion innovations, and they did so by creating new career paths along which younger officers specializing in the new tasks could be promoted. This analysis can be used to explain three successful cases of innovation in the U.S. Navy, Marine Corps, and Army, and also to explain the failed innovations initiated by the British navy between the world wars and by American army in its limited war in Vietnam.

From the Battlefleet to the Carrier Task Force in the U.S. Navy

The story of Rear Admiral William Moffett and the strategy he employed to make carrier aviation a reality in the U.S. Navy appears to fit the pattern suggested by the view of military organizations as political communities. Moffett was not an aviator but a successful battleship captain when he was appointed head of the Bureau of Aeronautics. He was selected by the senior leadership of the navy and the choice caused some consternation among navy aviators, who felt that only another aviator could advance their cause. One such described the process of selecting Moffett: "We [aviators] regarded the newcomer, Admiral Moffett, in our special aviation preserve somewhat critically, and asked our seniors how he happened to be chosen as the first Chief of Bureau. . . . They specified, patiently and in detail, the qualities of rank, presence, judgment, . . . and last but not by any means least, political aptitude that were required. Having set forth these specifications, these men and their seniors, in cold calculation, scanned the naval register and concluded that Moffett fit the bill."[1]

One of Moffett's principal allies, Admiral Joseph Reeves, was also a battleship admiral. Moffett, through a board headed up by his assistant chief of bureau, expanded the base of support for aviation by drawing into the ranks of the aviators officers too old to begin again in flight school. Navy captains were given the opportunity to gain their aviation wings as observers without going through the full training program and were then placed in aviation commands, eventually to be replaced by younger officers who had risen up from the bottom as aviators.[2]

But inserting young aviators into the promotion system was politically difficult. First, there were too many of them. Aviators were officers, not enlisted men. It was expected that many more aviators than surface ship officers would be killed in war, and so many more officers qualified to fly airplanes were needed in peacetime. But this threatened the balance of power between aviators and nonaviators within the navy.[3] The problem was ameliorated by a system in which aviators could go into the Navy Reserve in peacetime, to be activated in war. Second, as the strategic potential of aviation became better and better understood, aviators were less willing to subordinate themselves to officers who did not understand aviation and who used them in ways not consistent with the reformulation of naval warfare made possible by carrier aviation. The demand of aviators that they be commanded by other aviators drew a sharp

[1]Edward Arpee, *From Frigates to Flat-Tops* (privately printed, 1953), pp. 86–87.
[2]Archibald D. Turnbull and Clifford Lord, *History of United States Naval Aviation* (New Haven: Yale University Press, 1949; rpt., New York: Arno Press, 1972), pp. 245-46.
[3]"Policy on Naval Aviation Personnel," 26 September 1925, Serial No. 1298, File No. 421, Records of the General Board, National Archives, Washington, D.C.

reaction from traditional naval officers. The commander in chief of the U.S. Fleet wrote to the secretary of the navy that the aviators were "a menace to the service in their claim to the right to select who their commanding officers shall be." He was furious with the intervention in the personnel process by which Moffett ensured that only aviators were selected to command naval air stations and aircraft carriers. Moffett's responsibilities as bureau chief, the irate admiral argued, were confined to procurement, not assignments to command. Moffett replied to the General Board that "in the present stage of aeronautical development within the Navy it is impossible widely or sharply to separate the material and personnel functions. This perhaps is the crux of the matter. The older established order of things can not always be applied to the new art of Aeronautics."[4]

Moffett was challenging the entire political structure of the navy by introducing new cohorts of officers into the promotion system and by challenging the existing procedures for promotion. Ironically, he was also opposed by the aviators themselves. Their preferred solution was to create a separate flying corps in which they would be free from the oversight of nonaviators. Moffett, however, pointed out to them that the separate engineering corps that had existed within the navy in the nineteenth century had removed engineers from the promotion path that made them eligible to command ships and had badly fractured the officer corps at a time when it was necessary to integrate officers with engineering skills into the command structure of the navy, not segregate them into a separate class. By pursuing a separate flying corps, or a merger into a separate air service, the aviators were ensuring that they would never obtain the necessary sailing skills that would make them eligible to command aircraft carriers and to become admirals in charge of fleets incorporating aircraft carriers. Moffett's strategy was to resist the efforts both to keep aviators permanently subordinate to nonaviators and to have the aviation community secede from the navy. He continually argued that naval aviation should not be considered an end in itself but only an instrument of naval power. He testified before Congress that "supremacy in the air is of no use to anybody except as it affects conditions on the surface beneath."[5] As a result of his testimony before the Morrow Board, a congressional committee set up to investigate charges that the navy was paying insufficient attention to airpower, Congress declined to set up a separate air corps

[4]Arpee, *Frigates to Flat-Tops*, pp. 114–15, 118–19.
[5]Ibid., p. 123.

including naval aviation and in so doing protected Moffett from some of his external critics.

Thus aviators were kept within the navy, and surface-ship officers were drawn into aviation as observers. Both groups were available for the development of naval strategy and tactics based on a sound understanding of the potential of carrier aviation, and both became eligible for promotion to senior ranks. They also were in a position to influence the navy's ship-building programs. When France was defeated by German armored invasion, Congress authorized a vastly increased ship-building program, and the navy used this money to begin work on ten large aircraft carriers.[6] A naval aviation observer, Ernest J. King, who had commanded an aircraft carrier, became commander in chief of the U.S. Navy in World War II. In July 1941, King wrote to the General Board of his construction priorities. Fast carriers and slower, smaller "jeep" carriers were near the top of his list, exceeded in importance only by submarines for war against Japan and destroyers for escort duty against German submarines. Carrier construction was given clear priority over battleship construction.[7]

Another benefit of keeping aviators in the navy's promotion chain can be seen by tracking the men who made carrier doctrine. In 1931, as one aviation admiral later wrote, "The task of developing fleet carrier doctrine fell to Commander Aircraft Squadrons Battle Force, Rear Admiral Harry E. Yarnell, and the men around him. He equipped himself with some of the brightest aviators in the Navy— Captain Jack Towers, his chief of staff, plus Lieutenant Commanders Arthur Radford, Forrest P. Sherman, and Ralph Davison."[8] By World War II Jack Towers had become deputy commander in chief, Pacific. Forrest Sherman became the Pacific Command's assistant chief of staff for plans and after the war became chief of Naval Operations, as did Arthur Radford.[9] In these high positions they could influence the refinement of carrier warfare doctrine.

The last step toward the full exploitation of carrier aircraft was to create the doctrine of multicarrier task forces. If carriers were to be regarded as mobile airfields, it made sense to concentrate them at sea, to mass their offensive and defensive aircraft. This development

[6]Clark Reynolds, *The Fast Carriers: The Forging of an Air Navy* (Huntington, N.Y.: Robert Krieger, 1978), pp. 19, 38.

[7]King to Chairman, General Board, 30 July 1941, File No. 420-2, microfilm collection, Historical Collection, Naval War College, Newport, R.I.

[8]J. J. Clark and Clark Reynolds, *Carrier Admiral* (New York: McKay 1967), p. 44.

[9]Reynolds, *The Fast Carriers*, pp. 120–21.

was opposed by navy captains who regarded carriers primarily as ships, which ought to be dispersed in order to minimize their vulnerability to air attack. This opposition was strong, and Admiral Frederick Sherman could complain to his diary in 1942 and 1943: "The Navy high command in my opinion shows . . . no proper conception of handling carriers. We have yet to have a permanent two or more carrier task force trained to operate together. I seem to be alone in wanting to operate two carriers together."[10] Yet Sherman was one of the class of captains who commanded a carrier at the beginning of the war and who were promoted to flag rank after the first battles. By keeping aviators in the navy in positions of command, Moffett had ensured that they were in position to take advantage of rapid wartime promotions. After these officers had been promoted and some wartime analysis and experimentation had been performed, their ideas triumphed. Multicarrier task forces were adopted as part of formal navy doctrine in June 1943 with the publication of PAC-10 as the basic doctrinal document for carrier operations in the Pacific.[11]

Moffett died in 1933 and did not live to see the ultimate success of his strategy for innovation. His strategy was a slow one, and the pace of its success was linked to the pace of promotions. The beginning of this peacetime innovation can be dated to 1919, when the General Board first met to consider aircraft carrier aviation, or to 1921, when Moffett became chief of the Bureau of Aeronautics. Its culmination came in 1943 with the publication of PAC-10. The entire process thus took about twenty five years, or roughly the amount of time it took for a navy lieutenant to rise to the rank of captain or rear admiral. It was a strategy based on shaping the process of generational change in the officer corps, and as such, must have appeared maddeningly slow to the young officers advocating aviation, but it worked.

From Small Wars to Amphibious Assault in the USMC

Making things happen in the area of amphibious warfare was also a lengthy business. Official interest in the subject of advanced bases

[10]Wartime Diary of Admiral Frederick C. Sherman, Typescript of handwritten ms., n.d., Operational Archives, Naval Historical Center, Washington Naval Yard, Washington, D.C.

[11]"Current Tactical Orders and Doctrine: U.S. Pacific Fleet," PAC-10, Commander in Chief, Pacific Fleet, June 1943, Operational Archives, Naval Historical Center, Washington, D.C., p. iv–5, p. 6.

was first shown in 1905. The first full-scale amphibious warfare exercises took place in 1940, and the first combat assault took place in 1943. The process of peacetime innovation appears in this case also to have taken at least a full generation. The time required for the successful implementation of this innovation may, however, be explained by the large gap between the intellectual breakthroughs made by Major Ellis and the initiation of a bureaucratic strategy concerning the Marine Corps's promotion process. Ellis's brilliant plans, even though endorsed by the head of the Marine Corps, initially had remarkably little impact on the way the corps did business. The testimony of the most famous amphibious force commander of World War II, General Holland "Howlin' Mad" Smith, is eloquent on the subject of the lack of innovation in the corps in the early 1920s. Smith arrived at the Naval War College in 1920 to find not only disdain for the marines on the part of the senior Navy leadership of the college but also no sign that the problems of amphibious warfare were being seriously studied or taught: "Under the old Navy doctrine, a landing was a simple and haphazard affair, involving no planning and very little preparation. Assault forces were stowed in boats 5,000 yards off the beach and given a pat on the back. . . . Warships threw a few shells into the beach and that was all. Nobody took these landings seriously because the mere appearance of a large naval force off shore was supposed to inactivate the enemy. . . . Even the bitter lessons the British learned at Gallipoli had little effect on the War College."[12]

Smith was appalled to find the same "outmoded military thought as I had found at the Naval War College" when he arrived at the Marine Corps Schools in 1926 to take his field officer's course. An exercise in 1932 revealed to him "our total lack of equipment for such an undertaking, our inadequate training, and the lack of coordination between the assault forces and the simulated naval gunfire and air protection."[13] This lack of enthusiasm for the new form of warfare is noteworthy also for its occurrence during the Depression: in an era of scant resources, the Marine Corps might have been touting this new mission internally and externally in order to drum up more money for itself.

The intellectual breakthrough of Ellis had not been matched by a strategy to overcome normal bureaucratic inertia by transforming

[12]Holland Smith and Percy Finch, *Coral and Brass* (New York: Scribner's, 1949), pp. 48–49.
[13]Ibid., p. 57.

the officer corps. General Lejeune, while commandant, did not try to reorganize the Marine Corps around amphibious warfare. Although clear-sighted studies within the Marine Corps investigated the new critical military tasks identified by Ellis,[14] nothing was done to provide a permanent structure that would allow Marines to study and practice amphibious warfare in a sustained manner without prejudice to their chances for promotion. Small wars in Central America and China remained the principal practical occupation of the marines during the 1920s and 1930s. The Marine Corps Schools, where amphibious warfare was studied, were occasionally shut down so that students and faculty could be made available to serve in Nicaragua and China. Commandant Lejeune gave the whole subject of advanced-base forces two paragraphs in his memoirs, published in 1930, and made no mention at all of amphibious assault.[15] The secret war plans for amphibious warfare remained on the books, and perhaps security precautions made it impossible for Lejeune to discuss aspects of them in public. Be that as it may, the lack of discussion created a problem in implementing innovation. As one Marine Corps officer subsequently involved in amphibious warfare planning said, "No plan stuck away in the secret archives of the Navy Department, however excellent it may be, can be expected to function efficiently when the functioneers remain uninformed about the subject matter of the plans. What to do and how to do it must be a matter of common knowledge."[16]

It was therefore left to a subsequent commandant, General John Russell, to do for the marines what Moffett was doing for the navy, which was to fight the political battles for the careers of officers involved in the innovation so that the intellectual breakthroughs could be translated into real changes in capabilities. In the years up to and including 1934, he undertook a strategy with three components.

1. The most important aspect of this strategy was effected when Russell persuaded the chief of Naval Operations to establish the Fleet Marine Force (FMF). In Holland Smith's words, the creation of

---

[14]See, for example, "Naval Gunfire in Support of a Landing," a lecture given at Marine Corps Schools by Lt. Commander Albert Schrader, USN, 10 May 1929 (Historical Amphibious File, Breckinridge Library, Quantico, Va.), which explains how all the normal practices of naval artillery would have to be turned upside down to provide effective gunfire support for amphibious landings.

[15]John A. Lejeune, *The Reminiscences of a Marine* (Philadelphia: Dorrance, 1930), pp. 202, 445–48.

[16]Colonel Ellis B. Miller, USMC, "The Marine Corps: In support of the Fleet," 1 June 1935, p. 25, based on lectures delivered by Miller at the Marine Corps Schools during 1933, available from Breckinridge Library, Quantico, Va.

the FMF in 1933 "was the most important advance in the history of the Marine Corps, for it firmly established the Marine Corps as part of the organization of the U.S. Fleet, available for operations with the Fleet ashore or afloat." Instead of being frequently jerked away to fight small wars, the dominant component of the corps would henceforth be an integral part of the fleet as it conducted operations in distant waters. The practical result was that "for the first time, a permanent organization for the study and practice of amphibious warfare" was created.[17]

2. In the same year Russell directed that the Marine Corps Schools devote themselves exclusively to preparing a manual to train officers in the new methods of amphibious assault. The product was, in some ways, a reinvention of Ellis's wheel. The marines re-examined the old assumptions about amphibious warfare. Those assumptions, which had governed British amphibious warfare and which were reinforced in British minds by the unfortunate experiences at Gallipoli, were that troops landing on a beach could not survive if they faced prepared defenders possessing modern weapons. The British troops who landed at Cape Helles at Gallipoli had encountered Turks armed with rifles who cut the assault force down; the invaders' blood literally turned the waters red.[18] Amphibious landings had to avoid prepared defenses, utilizing either surprise, so that the defenders could not prepare defenses at the landing site, or the cover of night. Like Ellis, the officers at the Marine Corps Schools discovered that some island targets were so small that there was no hope of using surprise to find an undefended beach, and that operations large enough to seize prospective targets involved landing forces too sizable to be manageable at night. The Marines could not learn from British amphibious experience but had to reject it in favor of a completely different style of war. Intense naval gunfire and naval air strikes would be used to suppress defenses in the landing zone. Fast-paced assaults would then follow quickly behind the gunfire to keep pressure on the defenders and to capitalize on the effects of the preliminary bombardment. The assault would proceed rapidly, bypassing any strong points, so that it did not get bogged down by the defenders. Thus defensive firepower would not be avoided. Shock and speed would be used to destroy or overrun it. This was a revolution in

---

[17]Smith and Finch, *Coral and Brass*, p. 60.
[18]See the section on failures in wartime learning in chapter 4.

[83]

amphibious warfare, and it was preceded by the realization that in their efforts to learn from past operations, the marines had been, as one officer put it bluntly, "ON THE WRONG TRACK."[19]

3. Russell lobbied vigorously for the Selection Bill of 1934, which allowed the corps to retire overage officers, who had been promoted by virtue of their seniority, to make room for younger officers who were favorable to the new way of war. Holland Smith himself benefited from this law, becoming director of operations and training for the corps.[20]

Taken together, these components constituted a strategy for creating a new career track in which marines could develop the innovation, in which there was room for promotion and protection against interference. Although previously individuals had been able to study amphibious warfare and to advance an analysis of its problems, Russell's measures made it possible for the Marine Corps as an organization to move into the field of amphibious assault. The work at the schools resulted in a coherent doctrine for the development of amphibious assault, as outlined in the *Tentative Manual for Landing Operations*, first published in 1934 and reissued in revised form as the official U.S. Navy publication on the subject in 1938. Peacetime exercises were held every year up to 1942 to test and refine the concepts in the manual. Designs for specialized landing craft were tested. At Marine Corps insistence, landing craft and amphibious tractors were adapted from civilian designs when these proved superior to the products of the navy's design bureaus. The contracts for the amphibious tractors, in particular, were issued in 1940, and they were thus available in quantity in 1943 for the first amphibious assault on Tarawa, where they played a crucial role.[21]

The record of the corps in this regard was not perfect. The designs for large ships that transported landing craft, for example, had to be adapted from the British during the war, and there was a serious shortage of these ships that created bottlenecks that lasted almost until the end of the war. Equally serious were the deficiencies in air and naval gunfire support at Tarawa. It had not been fully appreciated just how accurate and prolonged this fire support would have to be if it were to root out dug-in Japanese defenders on the island. Neither aviators nor naval artillery men had practiced this kind of pinpoint bombardment, and so many Japanese soldiers survived the

---

[19]Ellis Miller, "The Marine Corps," p. 54; capitals in the original.
[20]Smith and Finch, *Coral and Brass*, p. 62.
[21]Jeter Isely and Philip Crowl, *The U.S. Marines and Amphibious War: Its Theory and Practice in the Pacific* (Princeton: Princeton University Press, 1951), pp. 36, 45, 68–69.

preliminary bombardment to kill marines as they waded ashore. But the doctrine was in place; officers understood the requirement for effective fire support even if they had not worked out the implementation. The problems at Tarawa could be treated as an indication of the need to improve the efficiency of existing forces, not as a challenge to the legitimacy of the existing doctrine and leadership. Intense, institutionalized criticism of existing gunfire practices were made, and training programs were quickly implemented to reform naval gunfire support. Although the proper use of gunfire remained a contentious issue between the marines and the navy, performance improved greatly in the next amphibious operations at Kwajalein and Roi-Namur.[22] Peacetime innovation had been successful enough that offensive amphibious operations were possible once the Japanese offensive thrusts had been stopped at Midway and Guadalcanal.

BUILDING THE AIRMOBILE DIVISION IN THE U.S. ARMY

The process of clearing the way for the development of helicopter aviation in the U.S. Army appears on the surface to have been quite different from the politics of organizational change seen in the U.S. Navy and Marine Corps. Previous accounts of this innovation have emphasized the critical importance of civilian intervention in overcoming resistance to helicopter airmobility within the military. Secretary of Defense Robert McNamara, in particular, and the civilians in the Systems Analysis division of the comptroller's office in the Office of the Secretary of Defense are seen as advocates of helicopter airmobility. Two biting memos sent to the army staff and to the secretary of the army by McNamara on 19 April 1962 have often been cited, even by army officers, as examples of the way in which civilian intervention can overcome the obstacles created by conservative officers and the tendency to protect resources devoted to traditional

---

[22]See, for example, the comments made after Tarawa by the commanding officer of the *USS Russell* and the supporting remarks made by the Office of the Chief of Naval Operations, in Secret Information Bulletin no. 15, *Battle Experience: Supporting Operations before and during the Occupation of the Gilbert Islands, November 1943*, issued 15 July 1944, Office of the Chief of Naval Operations, available at the Naval Historical Center, Operational Archives, Washington Naval Yard, and Naval War College, Newport, R.I., see also Isely and Crowl, *U.S. Marines And Amphibious Warfare*, pp. 234, 248, 251, 262, 268.

missions.[23] In the first of these memos, Secretary McNamara begins with a blast at the lack of progress within the army: "I have not been satisfied with Army program submissions for tactical mobility. I do not believe the Army has fully explored the revolutionary opportunities offered by aeronautical technology for making a revolutionary break with traditional surface mobility means. Air vehicles operating close to, but above the ground appear to me to offer the possibility of a quantum increase in effectiveness."

Harsh as this may seem, it was not unheard of for a secretary to express his desire to see more progress in a certain area. What follows was, however, very unusual in its attempt to tell the army not only what the secretary wanted, but how the army should do its business. "I therefore believe that the Army's reexamination of its aviation requirements should be a bold 'new look' at land warfare mobility. It should be conducted in an atmosphere divorced from its traditional viewpoints and past policies. . . . It also requires that bold, new ideas which the task force may recommend be protected from veto or dilution by conservative staff review."[24] McNamara further circumscribed army options by specifying the individuals who would serve on the task force that would review army airmobility requirements, naming well-known advocates of helicopters, beginning with General Howze as the board's president. The Howze Board recommended a greatly expanded role for helicopters in the army. It recommended the creation of airmobile divisions, which would use helicopters to transport infantry and artillery into positions where they could assault the enemy, and air cavalry brigades, in which helicopters armed with rockets and guns would seek out and attack enemy units. A more clear-cut case of successful civilian intervention to initiate or at least accelerate military innovation would, it seems, be hard to imagine. This instance tests the pattern of innovation suggested by the picture of military organizations sketched out in chapter 1 and exemplified by the cases discussed above.

An examination of the process by which ideas about helicopter aviation were transformed into new operational practices is further

[23]See, for example, Alain C. Enthoven and K. Wayne Smith, *How Much Is Enough: Shaping the Defense Program 1961–1969* (New York: Harper Colophon, 1971), pp. 100–104; John R. Gavin, *Air Assault: The Development of Airmobile Warfare* (New York: Hawthorne, 1969), pp. 275–76; John J. Tolson, *Vietnam Studies: Airmobility 1961–1971* (Washington, D.C.: GPO, 1973), pp. 18–19.

[24]Secretary of Defense to Mr. Stahr, 19 April 1962, reproduced as an appendix to *Howze Board*.

complicated by the fact that the events span a period characterized by both peace and war. While the U.S. Army was developing airmobility for its own use, it was also operating helicopter support for the army of the Republic of Vietnam. Beginning in 1961, and continuing with the deployment of the First Cavalry Division (Airmobile) to South Vietnam, experimental concepts of airmobility could be tested in battle. Data about combat operations involving helicopters became part of the development process. While much of this development took place before 1961, wartime experiments did play a role. Further, the exigencies of war undoubtedly played a role in accelerating the innovation. This case then, is worth careful study, to determine the extent to which civilian intervention can affect peacetime innovation, and to begin the examination of wartime innovation that will be continued in the next chapter.

The recognition of the need for airmobility and the development of the essential components of airmobile operations had taken place in the 1950s. A close examination of how the innovation was implemented indicates that much of the important work was completed in the same period, before the intervention of Secretary McNamara or the beginnings of the army's involvement in Vietnam. The events of this earlier period resemble closely those in the navy under William Moffett and those in the Marine Corps under John Russell. Indeed, there is testimony that army leaders deliberately modeled their strategy for bureaucratic change on that of Moffett. Beginning in 1955, there was a conscious effort on the part of a group of senior officers to restructure career paths in army aviation. Normal army personnel policies were bypassed in order to recruit into the aviation branch officers who simultaneously commanded the respect of the traditional leaders of the army, were on the "fast track" for promotion to senior rank, and were sympathetic to airmobility. They would eventually be replaced by young officers who would be trained as aviators but would also satisfy traditional army requirements for promotion to positions of command. This strategy closely resembles Moffett's use of naval aviation observers as promotable aviation advocates who would be succeeded by young naval aviators who would work their way up through naval aviation while also satisfying traditional navy requirements for promotion to positions of command. One of Howze's associates later stated that his superior "recognized, rightly, that the Army really needed to take a page, maybe at a lower level . . . from the navy. When the Navy went into Naval aviation . . . they . . . recognized that the thing to do was get the people on your team rather than fight them all the

[87]

way up. . . . [W]e needed to go out and recruit people who were already recognized."[25] Other historians have stated that in recruiting members of the traditional army elite into aviation, the airmobility advocates "followed the path of naval aviation." Howze himself was one of these recruits. Like Moffett, he was seen as part of the "mainline elite" of his service, an officer who "had a reputation for undramatically going by the book."[26]

In the army, this strategy was particularly necessary after World War II. Advocates of airmobility began with a large handicap. Army aviation had lost almost all of its combat veterans and combat capabilities when the U.S. Army Air Forces left to become an independent service. The remaining army aviators tended to be junior in officers without the combat experience that conferred credibility and promotability. In 1955, for example, only 4 percent of Army aviators had a rank above major.[27] The majority of aviators left in the army were pilots who simply liked flying. The only way they could do so after the departure of the Air Force was by piloting light planes on liaison missions transporting senior officers. These officers were not seen as combat leaders and commanded less respect than the officers in the combat arms. Howze commented that these pilots were "doing nothing but flying generals around." Any attempt to develop helicopter aviation using only these men would fail, as Howze noted, "simply because the Army wasn't about to take the ones who were pure aviators . . . and put them into points of high authority. It just isn't going to happen right now. . . . [W]e had to go outside the current aviation ranks." As is often the case, a literary work captures the sociology of an organization as well as the social sciences. In W. E. B. Griffin's book *The Brotherhood of War: The Majors*, there is the following exchange between an army officer and his wife in early 1955.

> "'I've had a battalion,' he said. "The brass trust people who have had commands. And they don't trust aviators. They don't take them seriously."
>
> "Is that fair?"
>
> "What's fair? It's the way things are. . . . I don't think the birdmen know, because they haven't been there, what the combat arms need,

---

[25]Interview with Brigadier General Glenn O. Goodhand, USA (ret.) by Colonel Bryce Kramer and Lieutenant Colonel Ronald K. Anderson, 9 May 1978. *Debriefing*, *CMH*, pp. 48–49.

[26]Frederic A. Bergerson, *The Army Gets an Air Force: Tactics of Insurgent Bureaucratic Politics* (Baltimore: Johns Hopkins Uniersity Press, 1980), pp. 104–6.

[27]Ibid., p. 102.

and what it takes to make a battalion work. And the birdmen don't think that we know, because we haven't been there, what aircraft are, and what they can do."

"So what happens?"

"Darwin. Survival of the fittest."[28]

The effort to promote airmobility in the army by reaching out to nonaviators who had established their legitimacy according to traditional army values began under Howze's predecessor at the Aviation Branch, Lieutenant Colonel Robert Williams. He had been exposed to the limited use of light aircraft and helicopters in the Korean war and had then gone on to work for General Gavin in G-3. Williams's first step was to approach Gavin to find a way to get West Point graduates and army officers who were the graduates of good colleges into the aviation program. This was not possible given army personnel policies, which required young lieutenants to hold tactical field assignments before going to flight school. Unfortunately, the tactical assignments took long enough to complete that the officers were overage for flight school by the time they were bureaucratically eligible. Williams proposed that Gavin change the policy, and, choosing a favorable moment, Gavin obtained the approval of the chief of staff of the army, General Maxwell Taylor, thus bypassing the officers normally responsible for personnel policy.[29]

Although it was important to begin the process by which young officers associated with aviation would reach senior rank, this strategy would have little short-term effect. The advocates of airmobility realized in 1954 that in the short term bureaucratic success depended, as one of the early aviators put it, on being able to "bring seasoned officers in general, primarily in the combat arms, into army aviation in order to make it go. . . . [T]he little aviation branch . . . were good men but they just hadn't had the experience in what makes the Army function."[30] Thus, just as the navy officers interested in promoting carrier aviation reached outside the aviation community to bring in a battleship admiral to head the Bureau of Aeronautics, so General Gavin reached out to bring in General

[28]Interview with Howze, *Debriefing*, CMH, pp. 14–15; W. E. B. Griffin, *The Brotherhood of War: The Majors* (New York: Jove Books, 1983), p. 139.

[29]Interview with Lt. General Robert Williams USA (ret.) conducted by George Pickett, May 1988, Northrop Analysis Center, Washington, D.C., pp. 2–4. I would like to thank George Pickett and General Williams for allowing me to make use of this interview material. Williams went on after retiring from the army to become the president of Bell Helicopter International.

[30]Interview with E. L. Powell, *Debriefing*, CMH, p. 27.

Howze, an armor officer, to be the first director of army aviation. Howze had the credibility that enabled him to reach out and draw respected nonaviators into aviation. He later recalled that he "said there should be another kind of career pattern based on the aviator who loves to fly . . . and wants to proceed with the development of Army aviation but wants to do it from the point of view of a combat soldier as distinct from being just a pilot. . . . [H]is ambition should be to become a Corps Commander, Army Commander." Like Moffett, Howze did not want aviators who wanted to dissociate themselves from the rest of the army but men who combined an interest in the innovation with a commitment to advancement within the larger organization. Howze also recalled: "In order to get some real enthusiasts, people who would associate their lives and progress in the Army with aviation, we had to go outside of the current aviation ranks. I selected many of those people myself."[31]

In addition to selecting Howze, Gavin himself began to select officers for entry into aviation, even though his formal responsibilities as head of G-3 did not include personnel decisions. In the period 1955–1957 he established special aviation classes for officers who were already lieutenant colonels and full colonels. Gavin selected the officers for these special classes using the same criteria the army used to select officers for promotion to the rank of general. This helped to insure that these lateral entrants into aviation would be promoted after they became pilots. This process of selecting potential flag-rank officers for training in aviation continued after Gavin left the military.[32]

By the end of the Eisenhower administration most of the intellectual problems involved in helicopter aviation had been solved and the key political struggles won. The technical problems of turning helicopters from vulnerable, unreliable aircraft into combat vehicles were solved by 1961. The XH-40 program was initiated in 1959 and yielded the UH-1, the "Huey" made famous in the Vietnam War, and its variant, the AH-1 "Cobra" helicopter gunship. The first army helicopters to utilize a turbine engine, they revolutionized the power and reliability of helicopters.[33] Army aviation was ready for a rapid takeoff in the 1960s because the key intellectual, political, and

---

[31]Interview with Howze, *Debriefing*, CMH, pp. 11, 15.

[32]Interview with Lt. Gen. Robert Williams, courtesy of George Pickett, pp. 5–6.

[33]George Pickett, "Airland Battle, Helicopters, and Tanks: Factors Influencing the Rate of Innovations," Northrop Analysis Center, Washington, D.C., p. 12; Tolson, *Airmobility*, p.7.

technological problems had been solved in the 1950s. As one aviator recalled: "Had the basic work not been done, you see, lead times in the development of aircraft and equipment are seven years at least. If you are going to procure them . . . by the time you . . . get the money and everything else, its probably four or five [more years]. So you just couldn't start from scratch."[34]

In this context, the McNamara memoranda to the army on helicopter aviation take on a different meaning. These memoranda did not initiate the innovation, but helped a group of senior officers win the endgame in their struggle to create combat units utilizing helicopters. What is equally important, it appears as if the McNamara memoranda were written by army aviation "mafia" and fed to the Office of the Secretary of Defense so that they would come back in a way that would help army aviation advocates. Late in 1961, Colonel Powell, who had worked with General Gavin on the airmobility problem for years, was working in the newly created Office of Defense Research and Engineering in the Office of the Secretary of Defense. Colonel Robert Williams had just been promoted to brigadier general and was assigned to command the Army Aviation School at Fort Rucker. Williams was approached by a civilian analyst in the System Analysis Office, Dieter Schwebbs, who told him that the army's justification for its aviation programs was inadequate. Williams went to the chief of Army Research and Development, Lieutenant General Trudeau and asked "to borrow Powell to work on the program memo [stating Army aviation program requirements] to make sure that what he (Schwebbs) told the Army to do was the right thing." Trudeau agreed, and while Powell was working Williams suggested that he insert the names of the army people who should be put on the board to review army aviation requirements. When he was through, Powell gave Schwebbs four documents. Schwebbs delivered them to the secretary of defense: "He took them to McNamara, and two of them were sent to the Army staff—and as I recollect it, they didn't change hardly a word. And one of those directives told in general what they (the Office of the Secretary of Defense) wanted done—and the other one, which was the one that made everybody mad, was that it told them (the Army) how to do it, including who should do it. And that's how it happened." General Williams only discovered the use to which Powell's memos had been put by the secretary of defense when he

[34]Interview with Goodhand, *Debriefing*, CMH, p. 49.

used the same memos as the basis for a speech and got into difficulty because it appeared he was plagiarizing from a sensitive internal document coming out of the secretary's office. Powell and
Williams kept their role in the affair secret for many years. The bad
blood between the army and McNamara meant that they would
have been very unpopular in the army if it became known how they
had secretly supplied McNamara with documents used to put pressure on the army.[35]

McNamara's office did play a role in the final stages of this innovation, but the larger contribution was the earlier creation of the officers who had already put in place the intellectual, human, and
technological programs that made helicopter aviation possible in the
early 1960s. Rather than as a clear case of civilian intervention to
force innovation this case should be understood as an example of
how senior officers won early struggles by satisfying traditional military requirements, and then used the civilian leadership to accelerate the final organizational changes. Seen in this way, the
development of helicopter airmobility more closely resembles the
pattern of innovation observed in the navy and Marine Corps than
superficial appearances would suggest.

What role did ongoing combat operations in Vietnam have on this
innovation? As early as December 1961, American helicopters with
U.S. Army pilots carried one thousand South Vietnamese paratroop-
ers into combat ten miles west of Saigon.[36] The Howze Board included an annex that discussed the results of a trip made to South
Vietnam in 1962 but made no detailed comments on helicopter warfare there. The annex merely concluded that conditions in South
Vietnam did not invalidate the idea of using helicopters to improve
tactical mobility. No insights concerning the use of helicopters in
combat were brought back from Vietnam for inclusion in the
annex.[37] The Howze Board recommendations led directly to the
formation of the Eleventh Air Assault Division commanded by
Brigadier General Harry Kinnard. This experimental division participated in field exercises and war games in the United States and
also sent six airmobile companies to Vietnam in 1964. The officers
from these companies came back to the Eleventh Division with
many valuable tactical lessons that refined existing concepts. The of-

[35]Interview with Powell, *Debriefing, CMH*, pp. 46–50; interview with Williams,
courtesy of George Pickett, pp. 7–9.
[36]Tolson, *Airmobility*, p. 3.
[37]Annex B, "Southeast Asia Trip Report," *Howze Board*, pp. 2–4, 9.

ficers from this division, including Kinnard, provided the nucleus of the first Cavalry Division that was formed and sent into combat in Vietnam in 1965.[38]

Once American helicopter units were involved in combat, the pattern of innovation appears quite different from that during the prewar period. In the cases of carrier aviation and amphibious assault, wartime experience served primarily to verify and refine prewar innovations. But the development of airmobility introduces a phenomenon that will be explored in greater depth in the chapters on wartime innovation. To some degree, wartime innovations in helicopter aviation did not represent the refinement of prewar concepts but rather an effort to completely rethink a military problem when peacetime concepts became inappropriate. Rethinking doctrine is a much more difficult task than simply learning how to better apply given concept to a particular circumstance and only the best, most intelligently commanded helicopter units in Vietnam were capable of this rethinking. An examination of those that did provides a partial introduction to the problems of wartime innovation.

At the organizational level, the heart of the airmobility concept was the integration of helicopters into air assault divisions and air cavalry brigades. Helicopter warfare in Vietnam was largely fought with helicopters that were *not* integrated into combat units.[39] More important, at the level of concepts of operation, it will be recalled that the peacetime army had emphasized the use of helicopters to keep large forces dispersed and then to quickly assemble them to launch large-scale attacks on the enemy. This was the concept which the first Cavalry Division employed in its first major action in Vietnam, when it came to the aid of South Vietnamese forces under attack at Plei Mei in the Central Highlands. The commander of the first Cavalry described this action as a test of prewar doctrine:

> [The] concept was to conduct an intensive search for the enemy. . . . By wide *dispersion*, made possible by excellent communications and helicopter lift, the Brigade was to sweep large areas systematically. . . . When contact was established a rapid reaction force was to be *assembled swiftly* and lifted by helicopters to strike the enemy. Rapid air movement of artillery batteries, plus extensive use of tactical air strikes, would provide fire support.

[38]Tolson, *Airmobility*, pp. 52–54.
[39]The 52d, 10th, 11th, 214th, and 269th Aviation Battalions were assigned on a *temporary* basis to support the 4th, 101st, 9th, and 25th Divisions in South Vietnam in 1966, for example. Tolson, *Airmobility*, 102–4.

Here was airmobility's acid test. The next few days would reveal whether three years of planning and testing would bear the fruits of victory—for a concept and a division.[40]

Intelligence reports after the battle indicated that the enemy regiment had judged its own casualties to be almost half of initial troop strength. The South Vietnamese force was rescued and 2,500 square kilometers of territory brought under temporary control. Kinnard's judgment was that the action had vindicated the work that had gone into airmobile warfare concepts of operation.[41]

Yet all was not well. The enemy was learning to counter the ability of American heliborne forces to mass infantry and firepower, preparing booby traps and ambushes for suspected landing zones and hugging close to American troops so that American airpower could not be brought fully to bear.[42] In battles in the first half of 1966 Kinnard tried to outwit the enemy using variations on the basic tactics of airmobility, choosing ever more unlikely landing zones for his troop-carrying helicopters in order to surprise the enemy.[43]

But later in that year, Kinnard's successor, Major General John Norton, noted a more fundamental change in enemy behavior. The North Vietnamese troops were less and less willing to engage large units of American helicopter borne troops. Smaller American "hunter-killer" teams had to be used to lure the enemy out so that he could then be engaged by airmobile units. Contact with the enemy was becoming more sporadic and combat less intense.[44] The first Cavalry launched Operations Thayer I and Irving in September and October of 1966, but the enemy would not fight, and broke up into small units to escape pursuit by large American units.[45]

With the enemy unwilling to fight the large units or face the firepower that the airmobile concept of operation was able to bring to bear, the first Cavalry no longer had a target against which it could

[40]Major General Harry W. O. Kinnard, first Cavalry Divison (Airmobile), "Combat Operations after Action Report: Plei Mei Campaign, 23 October–26 November 1965," dated 4 March 1965, CMH, p. 28; emphasis added.

[41]Ibid., Foreword, and pp. 14, 27–28, 70.

[42]Ibid., p. 14; Tolson, *Airmobility*, pp. 26–27, 49.

[43]Harry W. O. Kinnard, first Cavalry Division (Airmobile), "Operational Report on Lessons Learned 1/30–4/30 1966," 5 May 1966, submitted to the Assistant Chief of Staff for Force Development, CMH, pp. 2, 18, 50, 52.

[44]John Norton, first Cavalry Division (Airmobile), "Operational Report—Lessons Learned 1 May–31 July 1966," submitted to Assistant Chief of Staff for Force Development, 15 August 1966, CMH, pp. 11–12, 18–19.

[45]No author, Headquarters, second Brigade, first Cavalry (Airmobile), "Combat after Action Report: THAYER," 26 October 1966, CMH, p. 26.

use its special strengths. The local infrastructure of the Viet Cong, rather than enemy regular units, had to become the target. Refining airmobile tactics would not do the job. Abandoning these tactics now became the requirement. By the second half of 1966, the first Cavalry had to use its combat forces as infantry for patrols and ambushes. These nonairmobile units were better able to engage the enemy than the helicopter borne units. One-third of the first Cavalry's manpower, on foot, produced one-half of all contacts.[46] This pattern continued into early 1967, with long-range patrols by small units on foot and intensive local patrols producing the bulk of the contacts with the enemy.[47]

None of this discussion of the operations of the first Cavalry is meant to suggest failure on the part of that unit, nor should it be assumed that the unique mobility provided by helicopters did not continue to be useful well into 1967 and 1968, particularly in areas in which the enemy had not been exposed to airmobile units and had not learned how to counteract them or in areas such as the Mekong Delta, that were more easily patrolled by helicopters.[48] Other American units were not as successful as the first Cavalry in shifting away from large-scale assault tactics when they became less useful.[49]

Wartime innovation in helicopter aviation, then, proceeded along a path separate from peacetime innovation; it did not feed back into a process that refined the peacetime innovation. This suggests an important question that will be the focus of the next chapter: when wartime experience suggests that "going by the book" of peacetime procedures may not be satisfactory, how do military organizations go about the business of learning how to fight in a new way?

[46]John Norton, first Cavalry Divison (Airmobile), "Combat Operations after Action Report: Binh Dinh Pacification Campaign, Operations THAYER I, 13 September–24 October 1966, and IRVING, 2 October–24 October 1966," submitted to Commander US MACV, 13 January 1967, CMH, pp. 5, 40; Enclosure #10, p. 10–1.

[47]John Norton, first Cavalry Division (Airmobile), "Operational Lessons Learned, 1 November 1966–31 January 1967," submitted to Assistant Chief of Staff for Force Development, 15 February 1967, CMH, pp. 16, 18; Norton, "Operation THAYER II, 25 October 1966–12 February 1967," 25 June 1967, CMH, p. 36.

[48]See Tolson, *Airmobility*, p. 134, and the author's discussion of his own experiences as commander of the first Cavalry in 1967 in northern South Vietnam where only Marine Corps units had operated; Julian Ewell and Ira A. Hunt, *Vietnam Studies: Sharpening the Combat Edge: The Use of Analysis to Reinforce Military Judgement* (Washington, D.C.: GPO, 1974), pp. 75–95, 106–49.

[49]See first Infantry Division, "Operational Report—Lessons Learned, 1 August–31 October 1966," CMH, pp. 7, 10, 21–37, which simultaneously reports the relative ineffectiveness of large unit sweeps and recommends only minor changes in concepts of operation for combat.

Peacetime innovation is dependent at the intellectual level on an assessment of the security environment that leads to a perceived need for innovation which, in turn, leads to new concepts of military operations. At the practical level, it depends on a senior officer or a group of senior officers who first attract officers with solid traditional credentials to the innovation and then make it possible for younger officers to rise to positions of command while pursuing the innovation. If these propositions are true, then the absence of the specified factors should prevent an innovation from being successfully carried through, despite other sources of support. An examination of two noteworthy cases of aborted innovation suggests that this had, in fact, been the case. The Royal Navy pursued an independent role for carrier aviation during the 1920s and 1930s, only to see its carriers sunk without gain during World War II. The U.S. Army did pursue a counterinsurgency capability in the early 1960s, only to see large-unit infantry actions become the dominant form of warfare in the Vietnam War. The advocates of carrier aviation and counterinsurgency each had powerful sources of support outside their service but nonetheless failed, because of the absence of factors that are essential for peacetime innovation.

By 1918 the Royal Naval Air Service had approximately 3,500 airplanes and 55,000 officers and men in its service. A 200-plane air raid to be launched from Great Britain on the German port of Wilhelmshaven had reached the planning stages. By 1919 the Royal Navy had the first operational aircraft carriers in the world, and three additional ships were being converted to aircraft carriers.[50] The Royal Navy in the interwar period faced many of the same concerns in the Pacific as the United States—in particular, the need to develop sea-based airpower that could contribute to the defeat of the imperial Japanese Navy in war.[51] Like the U.S. Navy, the Royal Navy was pushed toward carrier construction by the limits on heavy cruisers imposed by the 1922 Washington Naval Treaty, and both navies circumvented these limits by converting two heavy cruisers

---

[50]Thomas C. Hone and Mark D. Mandeles, "Interwar Innovation in Three Navies: USN, RN, IJN," for the Office of Net Assessment, Office of the Secretary of Defense, MDA 903–82–C–0226, Washington, D.C., 1982, p. 2.

[51]Wesley Wark, "In Search of a Suitable Japan: British Naval Intelligence before the Second World War," *Intelligence and National Security* 1 (May 1986), 191.

into carriers.[52] By the beginning of the war in 1939, Great Britain had fourteen carriers built or in construction, versus eight for the United States and eleven for Japan.

Yet when naval warfare broke out between Germany and Great Britain in the waters off Norway in April 1940, the commander in chief of home waters, Admiral Sir Charles Forbes, left behind in port the one carrier in his force, the *Furious*. The captain of the *Furious* put out to sea anyway, on his own initiative, with eighteen torpedo bombers but no fighters on board, and he managed to get his ship sunk by gunfire from a German cruiser. The number of carrier-based aircraft the Royal Navy could bring to bear in Norway was so small that Winston Churchill was moved to inquire what was the point of building heavily armored aircraft carriers that could operate in harm's way if they carried so few aircraft as to make them valueless.[53] The Royal Navy, in short, was far from ready to exploit the potential of carrier-based aviation. Despite auspicious beginnings in World War I and impressive technological firsts, including armored decks, angled decks that made possible simultaneous aircraft launchings and landings, and optical aids for landing, the Royal Navy's efforts to develop carrier aviation between the wars have generally been regarded as seriously flawed. What had gone wrong? While aviation was supported by the Royal Navy, little effort was made to redefine the essence of naval operations taking carriers into account. There were no senior officers who drew officers from the traditional elite into the innovation and created new promotion paths for young aviators. Officers with combat aviation experience were compelled to leave the Navy, and young officers found themselves in an institutional environment in which they had no patrons who could advance their career. While some efforts were made to rectify this situation on the eve of war, by this time a genuine carrier aviation capability could not be created before the war was in its final years.

Like the U.S. Navy, the Royal Navy emerged from World War I with the idea that aircraft could serve as the eyes of the battleships. Unlike its counterpart, the Royal Navy retained this idea until 1938. Naval aircraft and aircraft carriers had no autonomous offensive or

[52]Stephen Roskill, *British Naval Policy between the Wars*, 2 vols. (London: Collins, 1968 and Annapolis: Naval Institute Press, 1976), 1:395. The United States converted the *Lexington* and *Saratoga*, the British the *Glorious* and *Courageous*.

[53]Geoffrey Till, *Air Power and the Royal Navy 1914–1945: A Historical Survey* (London: Jane's Publishing, 1979), pp. 13–16, 85, 90.

defensive missions.[54] This view of naval aviation was justified by the state of aviation technology in the early 1920s. But, unlike in the U.S. Navy, simulations were not used to explore what might be possible if aviation technology allowed improved aircraft performance. Without these simulations to suggest the importance of new technology, the Royal Navy was locked into a vicious circle in which assumptions about the ineffectiveness of naval aircraft perpetuated, and were perpetuated by, ineffective aircraft. Because it was believed that aircraft could not sink armored battleships, no money was made available for the development of carrier-borne bombers. Only general-purpose torpedo-spotter-reconnaissance aircraft were built and these, indeed, were not capable of independent offensive operations. Because it was thought that high-performance aircraft could not fly from the decks of aircraft carriers, it was believed that the air defense of carriers would have to be provided by antiaircraft artillery. Thus antiaircraft gunnery was developed, but not a high-performance fighter for use on carriers. No exercises were conducted to test what would happen if large numbers of fighters were available for the defense of carriers, so no effort was made to increase the number of aircraft that could be operated from a carrier. This attitude became so ingrained that fighters were unloaded from British carriers at the time of the Munich crisis and during the Norwegian campaign. When German fighters attacked the *Ark Royal*, which did have fighters on board, the fighters were kept below decks so that the antiaircraft guns would have a clear field of fire.[55]

Where William Moffett redefined aircraft carriers and insisted that they be thought of primarily as mobile airfields, not as ships, the Royal Navy took the opposite views. The assistant chief of Royal Naval Staff, Rear Admiral S. R. Bailey, commented on this attitude in 1934, writing that although the Royal Navy knew more about sailing carriers with the fleet "than anyone else does . . . we have been slow to arm our fleet with aircraft and to complete the carrier complements of aircraft and to become airminded." Nor is it surprising that as late as 1938 Royal Naval officers, on the basis of their experiences with carriers in exercises, tended to emphasize "the heavy liability a carrier may become."[56]

[54]Ibid., p. 139–141; Geoffrey Till, "Air Power and the Battleship in the 1920s," in Bryan Ranft, ed., *Technical Change and British Naval Policy 1860–1939* (London: Hodder and Stoughton), 1977, p. 112.
[55]Till, *Air Power and the Royal Navy*, pp. 143, 147–49; Roskill, *British Naval Policy*, 1:496–97, 2: 206–7.
[56] Till, *Air Power and the Royal Navy*, pp. 95, 164.

The leaders of the British navy also failed to win the political struggle to transform the officer corps to make it more aviation oriented. The senior leadership of the Royal Navy in the last year of World War I saw their overall professional credibility considerably damaged by their less than impressive handling of the German U-boat problem. The navy's management of the Royal Naval Air Service, moreover, had been flawed by some confusion of purpose and duplication of effort with the Royal Flying Corps. Royal Navy leaders, as a result, were in a weak position to resist proposals made by the committee set up by Prime Minister David Lloyd George to reform British military aviation policy. The committee, headed by Jan Smuts, in August 1917, proposed in essence to take aviation away from the army and navy and create a separate air ministry. Although there were legitimate arguments to be made against this reform, the navy's most senior officer, First Sea Lord Admiral Sir John Jellicoe did not make them. So bad were his position papers that the civilian First Lord of the Admiralty Sir Eric Geddes noted on one, "Better not use this argument." The Royal Navy was itself divided, and the commander in chief of the Grand Fleet, Admiral Beatty, endorsed the Smuts Committee report. The result was that 2,500 aircraft and 55,000 naval personnel were transferred to the Royal Air Force in April 1918.[57]

In this shuffle aviators with the most combat experience, seniority, and the best chances for promotion were removed from the navy. The contrast with the United States is striking. As a result of Moffett's strategy, the percentage of officers in the U.S. Navy who were aviators grew from less than 2 percent in 1916 to 11 percent in 1928.[58] The number of senior officers who were either aviators or observers grew as well. By 1927, the U.S. Navy had one operational aircraft carrier, and one vice admiral, three rear admirals, two captains, and sixty-three commanders receiving flight pay. The Royal Navy, in contrast, had six aircraft carriers in 1939, but only one flag-rank officer and few captains or commanders receiving flight pay.[59]

The organization that operated aircraft on Royal Navy carriers after the Smuts Committee reforms was a hybrid. The Fleet Air Arm was a mix of RAF aviators and RN officers temporarily transferred to the RAF for flight training. The twenty years after World War I witnessed a protracted struggle between the two services about how

[57]Roskill, *British Naval Policy*, 1:234–41.
[58]Turnbull and Lord, *History of U.S. Naval Aviation*, p. 265.
[59]Till, *Air Power and the Royal Navy*, pp. 45, 74.

the Navy officers in the Fleet Air Arm should be trained, who would command them when they were on ship or on land, who would decide how they would be used in war, and a host of other issues. Energy that might have gone into thinking through concepts of operations was dissipated in bureaucratic warfare.

At the same time, the organization of the Fleet Air Arm made it impossible for any Royal Navy officer interested in aviation to pursue that interest without jeopardizing his career as an officer. Any navy officer who wished to become an aviator had to be attached to the RAF for between three and four years. While with the RAF, it was not clear who would promote him. The RAF had little interest in promoting naval officers within its own ranks, since they would be returning to the Royal Navy. While with the RAF, however, they could not be promoted by the navy. Officers were assured that they would not lose their connection with the navy while with the RAF, but they discovered that when they did return they were seldom admitted into navy schools that would train them in traditional naval skills such as navigation and gunnery. Without these basic skills they had little chance of being promoted to the command of a ship. It is hardly surprising that the navy was never able to find enough volunteers to go to the RAF for training, nor that few officers with aviation backgrounds advanced to senior rank. Although the Fleet Air Arm was returned to the Royal Navy in 1938, the damage had been done. A vigorous wartime expansion program doubled the Fleet Air Arm's share of Royal Navy personnel, but many of the new recruits were low-ranking officers or enlisted men who had little influence on navy doctrine and policies.[60]

The attempt to create in the U.S. Army the capability to combat communist insurgencies provides one of the best documented cases of high-level civilian intervention to promote innovation. President John F. Kennedy involved himself personally in the effort to encourage the army to develop counterinsurgency (CI) capabilities in the early 1960s. The initiative received support at a critical juncture from the chief of staff of the army, General Harold Johnson. Yet, in the judgment of many army officers, the effort failed. It did not produce any significant new capabilities that could be used in the Vietnam war. Though the effort extended into a period of war, it can be argued that this was a case primarily of failed peacetime innovation, because the key failure was in not changing the way officers were

[60]Ibid., pp. 45–56; Roskill, *British Naval Policy*, 1:253, 1:257, 1:262, 1:358, 1:373, 1:394, 1:492, 1:494–95.

promoted after they left the war zone to return to the part of the army that was not involved in combat. It highlights the difficulty of creating new promotion pathways for officers with new capabilities, even with the support of senior leadership, when those new capabilities are not supported by well-established, if informal, patterns of promotion based on decades of organizational experience.

There is little doubt that President Kennedy placed great emphasis on the development of CI. The need to respond to what was perceived as a new communist strategy for the defeat of the West by means of "wars of national liberation" in poor, agrarian parts of the world was highlighted by Kennedy in his State of the Union address and in a special message to Congress in 1961. The White House issued a National Security Action Memorandum (NSAM 52) in May 1961 directing the military to reconsider its force posture for a possible military intervention against the communist insurgency in South Vietnam. In November 1961 Kennedy summoned to the Oval Office all of the major army commanders and told them that he, personally, wanted the army to develop new CI capabilities and that he understood that the army would not do so until it was convinced of the need. He set up a special group to oversee CI and selected its membership so as to lend it maximum political credibility. He included his brother, Attorney General Robert Kennedy, retired Chairman of the Joint Chiefs of Staff (JCS) General Maxwell Taylor, Deputy Secretary of Defense Roswell Gilpatrick, Chairman of the JCS General Lyman Lemnitzer, and Walt Rostow. This group was to supervise changes in army training programs aimed toward producing officers and enlisted men who had the special skills needed to fight effectively against guerrillas.[61]

Little resulted from the intense presidential pressure. In February 1965, the Army Concepts Development Center noted that it had yet to publish a handbook to train military advisors in CI tasks and skills. In May 1967, the Continental Army Command, which was responsible for the training of all Army troops in the mainland United States, noted that "there is insufficient doctrine on area warfare," using one of the terms used to refer to CI. Where the army had responded, it had done so by relabeling as CI programs existing doctrine for antipartisan warfare and rear-area security missions that had been developed to support army activities in a large-scale conventional war. Guard duty, map reading, and civil defense, all

[61]Andrew Krepinevich, *The Army and Vietnam* (Baltimore: Johns Hopkins University Press, 1986), pp. 30–32.

traditional army skills, were designated as CI tasks but in Army manuals were otherwise unchanged. The army added CI to the list of skills the Special Forces were to have. Broad studies of the politics, economies, and cultures of Southeast Asia, Greece, and Malaya were lumped together in a program to prepare army officers and civilians sent overseas to perform CI related jobs. These courses were of little use and were given poor marks by the students who were forced to take them.[62]

Thus, before 1964 planning at the operational level for war in South Vietnam remained focused on conventional large-unit tactics to defeat conventional invasions. The first army unit sent to Vietnam was an airborne brigade that had no significant CI training.[63] In fairness to the army, a complete reorientation of its training programs to create a genuine, army-wide capability for CI would have been difficult even with the best of intentions. Although the civilian leadership had decided that a structural shift had occurred in the strategic environment, the senior army leadership was far from convinced. The decisive wars of the future might be fought by and against guerrillas in the jungles and forests of the world, but the Soviet Union and its military remained a powerful force that could conquer the major industrial areas of Europe and Japan that had traditionally been at the core of American interests overseas. Had the security environment changed sufficiently that the major mission of the army should shift from the defense of these industrialized areas to the defense of poor, rural areas? Was the emphasis on counterinsurgency the result of a genuine change in the security environment, or was it only the personal fancy of one president that would be dropped by his successor? Units that were converted for CI would be less useful for the army's traditional missions, and the next president might decide they were not needed at all. It was reasonable for army leaders to consider what they would give up in pursuing counterinsurgency capability. The experiences of other military organizations clearly suggested that the army would have tremendous problems preparing for both a conventional war and for counterinsurgency. The skills and even personality traits officers need to be competent CI leaders are completely different from those required for the conduct of traditional land warfare. Patience is of

[62]Ibid., pp. 45–50; Douglas Blaufarb, *The Counterinsurgency Era* (New York: Free Press, 1977), pp. 71–73.
[63]See, for example, the discussion of the CINCPAC OPLAN 32–59 and CINCUSARPLAN 32–60, and the training of the 173rd Airborne Brigade in Krepinevich, *The Army and Vietnam*, pp. 80, 147.

greater value than aggressiveness. A tolerance of inaction, a sensitivity to local politics, and a predisposition to avoid the use of force are all necessary attributes of the successful practitioner of CI, and all are anathema to the successful practitioner of regular ground combat.[64] There is no doubt the army could have done more to prepare itself for CI in Vietnam, but it is also clear that the costs of doing so were higher than Kennedy seems to have realized.

Once the army was actually engaged in combat in South Vietnam, however, were there not new incentives to innovate and to develop genuine CI capabilities? If the army could not reconfigure itself to fight this kind of war, was there not an even greater need for American military advisors to train the South Vietnamese in CI strategies and tactics? The chief of staff of the U.S. Army during the critical years 1964–1968, General Harold Johnson, did, in fact, understand the importance of the CI mission in Vietnam and the desirability of training South Vietnamese forces in CI tactics. General Johnson was thwarted, however, in his more modest effort to develop CI capabilities by the strength of traditional beliefs about promotion pathways within the officer corps.

It is important to remember that the American military advisory presence in Vietnam remained relatively small throughout the war, never numbering more than about 750 officers and 1,500 enlisted men. Over 85 percent of the officers serving as advisors were self-described career officers who planned to serve in the army beyond the term of their initial contractual obligations. This compared with a level of only 48 percent self-described career officers in the army as a whole in 1971.[65] The motivations of the advisors were thus those of career officers who expected to serve long after the war in Vietnam was over.

Beginning no later than 1966, General Johnson came to understand that the war in Vietnam could not be won by the multibattalion search and destroy operations being conducted by General William Westmoreland unless those operations were complemented by a CI effort to pacify the countryside. He further understood that CI was best conducted by the South Vietnamese, but that they would benefit from the help of U.S. advisors skilled in CI strategies and tactics. This meant, in his judgment, that it was very important

[64]See the discussion in *U.S. Marine Corps Manual for Small Wars* (Washington, D.C.: GPO, 1940), chap. 1, sec. 3, pp. 11–18.
[65]Peter Miller Dawkins, "The United States Army and the 'Other' War in Vietnam: A Study of the Complexity of Implementing Organizational Change" (Ph.D. diss., Princeton University, 1979), pp. 54–55, 131.

to draw into the advisory mission some of the most capable officers in the U.S. Army, officers who had combat experience, were intelligent enough to learn from local circumstances, and who already had experience in Vietnam. Such officers were likely to be career officers. Johnson understood that in order to recruit men who had traditional army skills and credentials—men who could expect successful careers whether or not they involved themselves in the CI mission—service as an advisor had to be made something that would enhance their chances for promotion.

Promotion of an officer in the U.S. Army had always been accelerated by service as commander of a combat unit in war. No other kind of service conferred the same legitimacy and promotability. Advisory duty was not a combat command, and once the U.S. Army had combat units in action in Vietnam, service as an advisor was seen as second best in terms of career advancement. In a poll of former advisors, Peter Dawkins found that about 50 percent believed at the time of their assignments that advisory duty was detrimental to their careers. Roughly half concluded in retrospect, that their service as advisors had, in fact, prejudiced their chances for promotion.[66]

General Johnson was not unacquainted with the sociology of his army, and he attempted to overcome this problem by publishing a formal directive instructing all promotion boards to regard service as an advisor in South Vietnam as equivalent to a combat command in considerations for promotion. Army schools for midcareer officers were ordered to accept advisory duty as the equivalent of combat command when selecting officers for admission. But perception of the real, as opposed to formal, criteria of promotion were such that 68 percent of former advisors polled stated that they did not believe that promotion boards really treated advisory duty as the equivalent of combat command. They were strengthened in their skepticism by the fact that service as engineer, logistic officer, or security guard was also made the promotion equivalent of combat command thus diluting the value of advisory service. Promotion boards did, in fact, disregard General Johnson's directive in at least selected cases.[67]

Johnson persevered. He instituted a special program to recruit forty-four highly qualified officers to serve as "province senior advisors" in South Vietnam. Special benefits for the officers and their

[66]Ibid., pp. 68, 73; based on a poll of 509 former Army advisors.
[67]Ibid., p. 76–84; letter in possession of author disapproving service as an advisor as the equivalent of combat command dated 5 December 1968.

families were offered, and Johnson wrote personal letters to each asking them to participate in the program. Sixty-five percent of the officers approached turned Johnson down. He wrote them again. This time 59 percent turned him down.[68] Johnson had been unable to bring about a necessary innovation because he was unable to restructure the reality or perception of promotability within the service in the time he had available.

CONCLUSIONS

Peacetime innovations are possible, but the process is long. When changes in the structure of promotions that favor the innovation can be made, the officer corps changes over time, but the process is only as fast as the rate at which young officers rise to the top. Even in the most expeditious case of peacetime innovation observed—that involving helicopter aviation, at least eleven years from 1954 to 1965, were required for the idea of airmobile divisions to be turned into a functioning combat capability. A period of a generation or more was involved in the cases examined in the navy and Marine Corps. The pressures of the limited war in Vietnam did not seem to expedite the process of effecting an innovation.

But perhaps matters would be different in a total war, in which the nation and the military were committed to a struggle against a major military power and young officers could move more rapidly up the ranks because of openings created by combat. Perhaps increased opportunities to learn from combat experiences would speed up the process of military innovation. It is possible that wartime innovation is more effective in responding to changed military requirements than peacetime innovation. The chapters in the following section will explore these issues.

[68]Ibid., pp. 86–90.

WARTIME INNOVATION

[4]

The British Army and
the Tank,
1914–1918

Military innovation in peacetime is characterized, as I have shown, by uncertainties concerning the shape of future wars and by a process of sponsored or managed generational change within the officer corps. Although simulations of how future wars might be fought are possible and useful in initiating and refining innovations, certain knowledge of how a war will be fought and which new weapons or concepts of operations will be most effective is not possible. For this reason, a conclusive case for or against innovation is hard to make in peacetime. Innovation is slow because the turnover in the officer corps proceeds at a normal peacetime pace. Thus two inhibiting conditions can prevail: unsatisfactory older weapons and concepts can often linger because their inadequacy cannot be unambiguously demonstrated, and unsatisfactory officers who resist innovation can continue in service because they are not called upon to perform in battle.

But in wartime, surely, everything changes. Military innovations, defined in chapter 1 as major changes in the concepts of operation of a combat arm or the creation of a new concept of operations for a new combat arm, can be directly compared in battle with older ways of fighting. Combat provides, rather than speculative arguments, clear justifications for jettisoning old ways of fighting and adopting new ones. Old weapons and concepts that do not succeed in battle, it might reasonably be thought, will be shown up, and the need for their replacement will be obvious. In wartime, the military is "in business" in the same fashion as most bureaucracies are always so. In wartime, the military is performing the function it was designed to perform, which is fighting. Military innovation should be very different in wartime because of the possibility of organizational learning from ongoing operations and because of a second factor: in war, some officers die in battle, and others are promoted or relieved of command more rapidly than in peacetime, on the basis of their performance in battle. This changes the pressures and incentives facing officers who must consider innovation and may alter the ways in which officers with new approaches to fighting rise to positions in which their views affect the behavior of the organization.

In chapter 1, however, I presented the equally compelling argument that in wartime the learning that is relevant to innovation will be extremely difficult. War makes intelligence collection difficult. Moreover, the normal mechanisms for intelligence collection and analysis that are useful for organizational feedback and reform are not likely to lead to a recognition of the need for innovation. To recall the argument in chapter 1, if learning that leads to innovation is to take place, officers cannot simply ask themselves, "Are we executing our missions well?" Rather, they must ask: "Are our missions winning the war? Should we try to execute our missions better, or do we need to abandon them in favor of completely different missions and capabilities? How, in fact, can we measure the potential strategic effectiveness of an innovation?" This process of rethinking how operations lead to victory and devising new ways to measure how military capabilities relate to strategic success can be thought of as deriving a new *measure of strategic effectiveness*. Chapter 1 also suggested that because an innovation must be implemented in a very short period of time if it is to have an impact on the war, there is an advantage to tight central command, so that the organization can be rapidly reorganized once the need for innovation is perceived.

Traditional accounts of the development of the tank tend to emphasize the resistance to the new technology. This chapter will argue that the real problem was one of learning how to measure the effectiveness of the tank and how to use it most effectively. The British army was quick to perceive the technical merit of the tank, but it was slow to redefine strategic effectiveness and, because of its decentralized command, slow to reorganize. The measure of strategic effectiveness that confirmed the value of the tank was not "forward movement of the front line" but "the most efficient use of British manpower to kill German manpower." Because of intelligence collection and analysis problems this redefinition was not clearly made until 1918. Because of the lack of a strong central command, the British army was also slow to learn the most effective ways of using the tank.

In chapter 5, the special conditions under which wartime innovation was promoted rather than hindered by decentralized command will be explored. Submarine operations are, of necessity, decentralized, and in World War II turnover in the U.S. Navy's submarine officer corps did lead to rapid innovations in operating doctrine in the Pacific. The chapter will conclude, however, that the lack of centralized intelligence analysis hindered the emergence of a new measure of strategic effectiveness, and that this, in turn, hindered the assessment of the impact of the submarine war.

Chapter 6 will examine innovations in the forces involved in the American strategic bombing of Europe in light of the concept of redefining strategic measures of effectiveness. There was a clear effort in this case to develop new measures of strategic effectiveness. Yet wartime constraints on what could be learned from combat meant that the success of the innovations was greatly dependent on good fortune.

REDEFINING VICTORY: THE BRITISH ARMY IN WORLD WAR I

The British army entered World War I with a general incapacity to learn from experience that was noted by contemporary military observers. The military historian G. F. R. Henderson, a professor at the British Staff College, had served as the director of intelligence to General Roberts in the Boer War. In an essay written before World War I he pointed to the British military's lack of formal means to collect data and analyses from individual soldiers with combat experience and with these develop a coherent view of the changing character of modern warfare. British soldiers acting as observers "had carefully studied and minutely reported on" the contemporary military operations of foreign armies, and their data had revealed points that are now obvious, such as the importance of artillery bombardment of entrenched infantry before an assault, the suicidal character of assaults on prepared defenses, and the fallacy of believing that an attacking infantry line could with its own rifles suppress the defensive fire coming from soldiers with modern rifles firing from trenches. The problem was not that individuals had not collected and studied modern military conditions but rather the lack of institutional concern: "In the British army no means existed for collecting, much less analyzing the facts and phenomenon of the battlefield and the range. Experience was regarded as the private property of individuals, not as a public asset, to be applied to the benefit of the army as a whole. . . . [T]he suggestion that a branch should be established for the purpose of dealing with strategical and tactical problems involving both technical knowledge and patient study was howled down by the economists."[1] While improvements were made in the quality and number of intelligence officers between the time Henderson wrote his essay and the beginning of World War I, the problem was aggravated by a widely shared distaste for detailed analysis among British officers. Field Marshall Douglas Haig, for example, believed that such studies would bind the hands of the wartime commander in the field. What was equally harmful was the practice of disbanding the peacetime General Staff so that its officers could take up field commands once war had begun. There was also a lack of appreciation that analysis and planning would be needed just as much during the war as before it.[2] The result, according to the recollections of a chief of the Imperial General Staff, Field Marshall Lord Milne, was the absence of any coherent framework for analyzing the relative merits of alternative modes of fighting during the war: "You will remember what happened in the late war; we jumped from one conclusion after another and in the end I think you all came back to what you had been taught in the beginning—the use of the rifle."[3]

Intelligence functions were divided between the intelligence unit in the General Headquarters of the British Expeditionary Force in France, directed in 1916 and 1917 by Brigadier General John Charteris, and the director of military operations and intelligence for the Imperial General Staff in London after 1915, General Sir George Macdonogh. Serious differences arose between the two offices, since Charteris appeared to be systematically underestimating the weaknesses of the German Army. The differences became so great that at one point during the war the chief of the Imperial General Staff, Sir William Robertson, wrote to the commander of the British armies in the field, Douglas Haig, to say that he "cannot possibly agree with some of the statements" made in an intelligence appendix written

---

[1]G. F. R. Henderson, *The Science of War: A Collection of Essays and Lectures 1892–1903* (London: Longmans, Green, 1905), pp. 418–19.

[2]John Gooch, *The Plans of War: The General Staff and British Military Strategy 1900–1916* (New York: Wiley, 1974), pp. 118, 120; Timothy Travers, *The Killing Ground: The British Army, the Western Front, and the Emergence of Modern Warfare 1900–1918* (Boston: Allen and Unwin, 1987), p. 67; Hankey, *The Supreme Command 1914–1918* (London: Allen and Unwin, 1961), 1:232.

[3]Travers, *Killing Ground*, p. xix.

by Charteris, and that he, Robertson, would decline to have the report circulated to the War Cabinet.[4] These institutional arrangements did not substantially improve during the course of war.

New concepts for analyzing intelligence did emerge, however. In the decades before the war, there were many signs that individual officers in the British army had begun to understand the impact of the revolution in firepower on the balance between the offensive and defensive functions. Field Marshall Sir John Burgoyne wrote extensively about the lessons to be learned from the Crimean War and the American Civil War about the destructiveness of breech loading rifles and rifled artillery and the importance of field entrenchments. In the 1870s the pages of the *Journal of the Royal United Services Institution* contained many articles by lower-ranking officers with titles such as "The Amount of the Advantage Which the New Arms of Precision Give the Defense Over the Attack" and "Shelter Trenches or Temporary Cover for Troops in Position."[5] The first commandant of the Royal Staff College, Sir Patrick MacDougal, predicted on the basis of his observations of the American Civil War and the Franco-Prussian War that future wars would be decided by artillery and long-range rifle fire, and that mass infantry formations would have to be discarded on the battlefield. In his book *Modern Infantry Tactics* he left no doubt about the way these developments strengthened tactical defense: "A front held by good troops undemoralized is practically unassailable under the present conditions of fire." He predicted that future wars "will, in all probability, be decided by strategic rather than tactical maneuvering." The ideal concept of operations would be a defensive-offensive one that forced the enemy to wear himself out by attacking prepared defenses.[6] Lord Kitchener in 1898 noted the impact of modern firepower on Africans, and he "confessed that he was wondering what was going to happen when . . . [the British] had to undertake the offensive under similar conditions."[7] By 1907 Kitchener was arguing that infantry would not

[4]Cited in Victor Bonham-Carter, *The Strategy of Victory 1914–1918: The Life and Times of William Robertson* (New York: Holt, Rinehart, Winston, 1963), pp. 251, 258, 399 (note 22).

[5]Cited in Jay Luvaas, *The Education of an Army: British Military Thought 1815–1940* (Chicago: University of Chicago Press, 1964), pp. 90–91, 95.

[6]Ibid., pp. 111, 114.

[7]Cited in Ernest D. Swinton, *Eyewitness: Being a Personal Reminiscence of Certain Phases of the Great War, Including the Genesis of the Tank* (Garden City, N.Y.: Doubleday, Doran, 1933), p. 97.

be able to advance over open ground against modern artillery and rifle fire without the aid of darkness and entrenchments.[8]

But if individuals within the British officer corps did appreciate some of the tactical and operational consequences of the changes in military technology, the army as an institution was not able to prepare itself in peacetime so that it would be able either to fight well under the new conditions or to learn and innovate in wartime. If modern firepower made infantry assault and conventional offensives difficult, if not impossible, how should the British army begin to look for alternative methods of winning a European war? Learning from experience begins by defining a problem to be solved or identifying a measure of effectiveness with which alternatives could be judged. In 1912, the British General Staff argued that a general European war would last at least six months, and that, if it lasted longer, it was preferable to wage war by means of an economic blockade of Germany rather than by raising a mass army. A small expeditionary force plus the blockade would attack Germany's economic vulnerabilities, while the alternative of raising a large British army would disrupt the British economy.[9] The strategic measure of effectiveness defined by the economic theory of victory was the relative national economic damage suffered by the combatants, not battlefield success. This strategic perspective encouraged attention to the question of how the British should fight when the war proved long and the blockade unsuccessful. When this did occur, the British reaction was a general casting about for ideas. In his famous "Boxing Day" memorandum of December 1914, the secretary of the Imperial General Staff, Maurice Hankey, crystallized the general problem: "The remarkable deadlock which has occurred in the western theater of war invites consideration of the question whether some other outlet can be found for the effective employment of the great forces of which we shall be able to dispose in a few months time."[10]

---

[8]Shelford Bidwell and Dominick Graham, *Fire-Power: British Army Weapons and Theories of War 1904–1945* (London: Allen and Unwin, 1982), p. 31.

[9]For general British predictions of a short war, see for example, Gooch, *The Plans of War*, p. 120. The 1912 staff estimate is cited in David French, *British Strategy and War Aims, 1914–1918* (Boston: Allen and Unwin, 1986), p. 15; the expected economic impact of mass mobilization is discussed in David French, *British Economic and Strategic Planning 1905–1915* (London: Allen and Unwin, 1982), pp. 34–35, 55–64.

[10]The text of this memo is provided in many sources, including Martin S. Gilbert, *Winston S. Churchill*, companion volume 3, part 1, *July 1914–April 1915* (Boston: Houghton Mifflin, 1973), p. 337.

The problem was being posed for the first time, and Hankey suggested a number of ideas, ranging from use of the tank, to chemical warfare, to opening a new strategic front. How could intelligence be used to evaluate the alternatives relative to each other and to the present strategy of continued pressure on the central front, the course preferred by senior officers?

Lord Kitchener provided the definition of the problem that came to dominate. By 1909 he had concluded that an Anglo-German war would last at least three years and that it would likely not end before the military manpower of Germany was exhausted. From this premise, he laid out in August 1914 a plan for a 700,000-man army to be raised by April 1915 and a million additional volunteers by the end of that year. Kitchener based his policy on the assumption that the first years of the war would see the German, French, and Russian armies exhausting themselves against each other while his new armies were raised and trained. When the continental armies were on the verge of collapse, his New Armies would intervene to determine the outcome of the war.[11]

Kitchener had defined the problem as one of attrition. The strategic measure of merit was of the exchange ratio of soldiers lost on the battlefield. Whichever method produced the most favorable exchange ratio, either by conserving friendly manpower or by killing the maximum number of the enemy, was to be preferred. However, continued consideration of amphibious landings in northern Europe and the actual pursuit of the Dardanelles expedition show that Kitchener's concept was not accepted at once. Winston Churchill, for example, moved only slowly and painfully from a preferred strategy of amphibious landings on the periphery of Germany and Turkey, which would win the war by depriving Germany of her allies, to a reluctant acceptance of the strategy of killing Germans most efficiently.[12] Douglas Haig continued to search for a breakthrough on the Western Front that would lead to rapid forward movement of his troops, the traditional measure of success.[13] As long as separate, conflicting measures of strategic effectiveness persisted, no clear lessons could be learned from the war. Disagreement about what constituted success or failure at the strategic level led to disagreement about whether continued fighting in the trenches was a necessary but painful step toward winning or simply futile bloodletting.

[11]French, *British Strategy,* pp. 24–26.
[12]See, for example, Tuvia Ben Moshe, "Churchill's Strategic Conception during the First World War," *The Journal of Strategic Studies* 12 (March 1989), 5–21
[13]Bidwell and Graham, *Fire-Power,* pp. 71, 88, 99.

The intellectual trend was clearly in Kitchener's direction, but his view was slow to take hold. By June 1915, Lord Curzon addressed the War Cabinet in terms of killing Germans in the most efficient way, but his conclusion was that there was no preferred method for arriving at that goal: "The war seems to me to be resolving itself largely into a question of killing Germans. For this purpose, viewing the present methods and instruments of war, one man seems to me to be about the equivalent of another, and one life taken to involve another one. If then two million (or whatever figure) more of Germans have to be killed, at least a corresponding number of allied soldiers will have to be sacrificed to effect that object."[14]

In November of 1915, General William Robertson, who was then about to become chief of the Imperial General Staff, wrote a memo that Prime Minister Asquith circulated to the cabinet that acknowledged that the war might have to be won by exhausting German manpower, but he argued that the evidence did not suggest that the ratio of lives exchanged would favor the British if they went on the tactical defensive. Offensive action was the only way to force the enemy to fight and so to wear him down, however much civilians might not wish this to be so. Robertson asserted that attacks that penetrated the enemy lines or forced the enemy to counter attack to regain lost ground imposed heavier casualties on the "defender" than on the attacker.[15] The failure of the Gallipoli expedition at the end of 1916 ended the argument for a new strategic theater. It was not until April 1917 that Robertson and the General Staff had totally accepted the concept of winning the war by killing more Germans than the Germans killed Allied soldiers, with the civilian members of the War Cabinet emphasizing the role of attrition in bringing about a revolution in Germany.[16] Haig's hopes for a breakthrough in the offensives of the summer of 1917 suggest that he had not accepted the strategic concept of attrition even then.

Even if the senior field commander was not totally of the same mind, once the British military staff was focused on the problem of how to kill Germans most efficiently, it was in a position to learn from its experiences. In hindsight, what is most striking is how evenly the postwar evidence shows casualties to have been divided

[14]Travers, *Killing Ground*, p. 119.

[15]Memo cited in William Robertson, *Soldiers and Statesmen 1914–1918* (New York: Scribner's, 1926), 1:184, 1:196–200.

[16]John Gooch, "Soldiers, Strategy, and War Aims in Britain 1914–1918," in Barry Hunt and Adrian Preston, eds., *War Aims and Strategic Policy in the Great War, 1914–1918* (London: Croom Helm, 1977), p. 31; French, *British Strategy*, p. 23.

between offense and defense, supporting in part Robertson's assertion that the offensive tactics of the Allied armies were not more
wasteful of British lives than defensive tactics. Postwar data indicates that in the Battle of Artois of May 1915, the French, who were
on the attack, took 100,000 casualties, versus 75,000 for defending
Germans. At Verdun, the French were on the tactical offensive, trying to dislodge the Germans from a salient they had seized, and
they suffered 360,000 casualties, while the Germans lost 330,000. In
the entire five month battle of the Somme in 1916, the British and
French were on the attack and lost 590,000 casualties, the German
defenders 600,000. In their last offensive in the spring of 1918, the
German attackers lost 348,000 men, the Allied defenders 332,000.[17]
The data indicate, first, that it is hard to distinguish clearly between
"the offense" and "the defense" in these battles, since they generally involved attacks and counterattacks on both sides and, second,
that the casualties suffered by the nominal attackers and defenders
were rather equally balanced. If British generals had had the benefit
of this postwar data, it is not at all clear that the lesson they would
have taken from wartime experience was that defense was superior
to existing tactics, or that organizational learning would have favored innovation.

But, of course, British commanders did not have postwar data.
What is most striking in reviewing the intelligence available to them
is how little hard data they had with which to evaluate strategic options and innovations. At the very beginning of hostilities Prime
Minister Asquith wrote that "oddly enough, there is no authentic
war news—either by land or by sea: all that appears in the papers is
inventions."[18] Military intelligence sources seemed little better; in
particular, there were few hard figures about German casualties, the
intelligence component that was central to the measure of effectiveness proceeding from the goal of efficiently killing German soldiers.
In December 1914, Lord Kitchener asked the then director of military operations at the War Office, Charles Callwell, to estimate the
time at which the Germans would run out of men of military age
due to the casualties they were suffering. Callwell's reaction to this
request is revealed in a personal letter: "K has told me to prove that
the Germans will run out of men within the next few months—and

---

[17]Figures from Tony Ashworth, *Trench Warfare 1914–1918: The Live and Let Live System*
(New York: Holmes and Meir, 1980), pp. 50–56.
[18]Cited in David French, "Sir John French's Secret Service on the Western Front,
1914–1915," *The Journal of Strategic Studies* 7 (December 1984), 424.

I have. I could just as easily proved that they were good for another two years. One must be a mug indeed if one cannot prove anything with figures as counters."[19]

Writing after the war, Callwell explained his problem further: "The question seemed to base itself on what premises you thought fit to start from. You could no doubt calculate with some certainty upon the total number of Teuton males of fighting age being somewhere about fifteen million in August 1914, upon 700,000, or so, youths annually reaching the age of eighteen and upon Germany being obliged to have under arms continually some five million soldiers. After that you were handling indeterminate factors."

How many men did the Germans have? How many were being killed? The only available data were "based on suspicious enemy statistics, and the perplexities involved in the number of wounded—who would, and who would not be able to return to the ranks."[20] In order to answer Kitchener's request, Callwell made two basic assumptions, one about influxes to the German army and one about casualties. He assumed that the group of German men just reaching military age were not immediately available for combat, presumably because they were not yet trained, and that German casualties were running at a rate equal to two to three times French losses. At that rate, the Germans would run out of men in six months. The War Cabinet doubted this estimate, preferring an estimate that showed the Germans collapsing slightly later, toward the end of 1915. Callwell appears to have misjudged the intent behind Kitchener's request for analysis, since subsequent historians have shown that at this time Kitchener was the most realistic of all the senior leaders, sticking to his original estimate of a three-year war.[21] Nonetheless, Callwell's estimate of a German collapse in six months was retained, but rolled forward throughout the war, so that victory was always just six months off. George Macdonogh, Callwell's successor, continued to overestimate the size of German losses through 1915, and his military estimate, circulated in the Cabinet in support of a 1916 offensive, predicted the collapse of German manpower in that year. He subsequently underestimated German manpower reserves by continuing Callwell's practice of assuming that the youngest cohort of German men of military age would not be available for combat.[22]

[19]Ibid., p. 436.
[20]Major General Charles E. Callwell, *Experiences of a Dug-Out 1914–1918* (London: Constable, 1920), p. 109.
[21]French, *British Strategy*, p. 65.
[22]Ibid., pp. 161–63; Travers, *Killing Ground*, pp. 115–16.

These estimates continued to circulate in large part because of the absence of hard data to confirm or deny them. Human intelligence sources, when they were reestablished after the prewar networks had been disrupted by the onset of hostilities, could not, by their nature, be expected to give systematic, comprehensive reporting on German battle casualties. Communications intelligence derived from the decoding of enemy radio transmissions was available; the British and French armies had the ability to read German army radio transmissions and even field telephone communications.[23] But communications intelligence tends to be fragmentary and to reflect the honesty (or dishonesty) and concerns of the military commanders whose messages are being read. Commanders might or might not need to communicate the information their enemies would like to intercept. German army radio traffic was dominated by orders to move units and reports concerning their location and readiness. Early in the war, the British army often had remarkably precise information about the German army's order of battle, in terms of the number and location of military units stationed on the front, and less frequently they had warning of impending German attack. As the war progressed, a combination of British errors and German improvements in security reduced the value of this intelligence.[24]

Even when it was effective, signal intelligence did not prevent large uncertainties about German casualties. The director of intelligence for the British armies in the field in the period 1916–1917, John Charteris, owned all the signal intelligence assets, and his assessments of German losses were, in the words of one of his subordinates, slanted "to prove a given theory—exhaustion of German manpower."[25] Usually the only hard data available to British intelligence about German casualties were derived from intercepts of communications concerning the movement of enemy units to and from the line and the identities of the units to which captured German soldiers belonged. These formed the basis of estimates of German casualties, but the estimates nonetheless involved large, untested assumptions. The intelligence staff of GHQ wrote: "It is estimated that a division that is withdrawn exhausted will have suffered approximately 3,000 casualties in its infantry and 200 casualties in its

[23]See, for example, Marshall Joffre, *The Personal Memoirs of Joffre*, 2 vols., trans. T. B. Mott (New York: Harper, 1932), 1:222–24; David French, "French's Secret Service," pp. 433–34; John Ferris, "The British Army and Signals Intelligence in the Field during the First World War," *Intelligence and National Security* 3 (October 1988), 26.
[24]Ferris, "British Army Signals Intelligence," pp. 30, 34–35.
[25]Ibid., p. 33.

other services." Similarly, it was estimated that when a given German age group, for example, that of men reaching the age of twenty in 1918, provided 15 to 16 percent of the prisoners of war taken by the British, then that cohort was "exhausted as a means of replacing further casualties."[26] Such estimates were bound to be soft. The uncertainty concerning German casualties persisted even after the war, producing bitter debates among staff historians writing the official history of British military operations in World War I.[27]

If British leaders wished to discover not simply the total number of casualties suffered by the enemy but also how many of the casualties were produced by different weapons or by different types of military operations, the problems of data collection multiplied. It was presumably easier to physically distinguish casualties caused by chemical warfare from other types, since the wounds differed. Yet there appears to have been little reliable data available to either Allied or Entente powers about the effects of the initial use of gas in combat. The Germans printed that their first use of chemical weapons in April 1915 claimed two hundred hospitalized victims; the Allies gave a figure of twenty thousand wounded and killed. No doubt both sides had reasons to distort the data, and a historian of gas warfare has written that "neither claim is believable, nor will the facts ever be established."[28] If such uncertainty prevailed with regard to chemical warfare, one must appreciate the difficulty of sorting out casualties from rifle fire, machine gun fire, and artillery shrapnel.

For the British in World War I, then, learning from experience required the formulation of a question to be answered, so that it could be determined what it was they were trying to learn about. This task had been accomplished by 1917, when a measure of effective-

---

[26]21 August 1917, Note by GHQ Intelligence on "German Manpower, Casualties, and Morale," British War Office (WO) record group 106, piece no. 1514, Files of the Director of Military Operations and Intelligence/Reports submitted by the Director of Military Intelligence General Sir George Macdonogh), Public Records Office, Kew, United Kingdom (hereinafter cited as PRO followed by WO record group/piece number). See also the rough catergorization of German divisions into "tired" and "fresh" and the report on German age cohorts in the 1 September 1917 memo by John Charteris to Douglas Haig summarized in his diary in John Charteris, *At GHQ* (London: n.p., 1931), p. 249.

[27]David French, " 'Official but Not History?': Sir James Edmonds and the Official History of the Great War," *Journal of the Royal United Services Institute* (March 1986), 60–61.

[28]L. F. Haber, *The Poisonous Cloud: Chemical Warfare in the First World War* (Oxford: Oxford Univeristy Press, 1986), p. 39. Haber himself is the son of the German scientist and officer Fritz Haber, who was responsible for the development of German chemical warfare in World War I.

ness was agreed upon. It was decided that what the British in effect needed to do was to learn to destroy German manpower most efficiently. But then the task of learning was made very difficult by the problems of data collection. In the absence of reliable data, assumptions were made that impeded learning and innovation. German casualties were repeatedly estimated to be bringing the enemy to the point of collapse. Battlefield experience, viewed through the lens of British intelligence, indicated no need to learn and no need to innovate at the strategic level.

MANPOWER AND THE TANK

And yet there was a major innovation introduced by the British, the tank. If intelligence did not point to the need for innovation, did organizational learning and the measure of strategic effectiveness play any role in the introduction of the tank? The genesis of the idea of the tank in France and Great Britain and its initial rebuff by the British army has been described in many accounts.[29] It is clear from the standard accounts that the idea of putting a gun in an armored, self-propelled carriage on caterpillar treads occurred independently to many people at roughly the same time. What is in question here, however, is not the technological developments of 1914, but the behavior of the British army and how it changed. Given the ways in which it measured military effectiveness, how did the British evaluate the tank as an alternative to its existing ways of fighting?

From its inception, the tank received the support of the highest ranking officers in the British army. The idea was supported by Sir John French in June 1915 while he was commander in chief of the British Expeditionary Force on the grounds that it had "considerable tactical value."[30] French's successor, Douglas Haig, inquired as to the progress in developing the "Caterpillar Machine Gun Destroyers or 'Land Cruisers'" in December 1915. He wrote after the first proving-grounds tests with production models that the tank "can usefully be employed in offensive operations by the Armies under

---

[29]See, for example, B. H. Liddell Hart, *The Tanks, 1914–1939*, 2 vols. (New York: Frederick Praeger, 1959), 1:17; Swinton, *Eyewitness*, pp. 11, 81, 84, 98, 148–49, 161–63; Winston S. Churchill, *The World Crisis 1911–1918*, 2 vols. (London: Odhams Press, 1938), 1:510–14; chapter 13 of Robin Prior's *Churchill's 'World Crisis' as History* (London: Croom Helm, 1983) adds some interesting detail to the original accounts but does not succeed in overturning the basic picture of Churchill's role in the genesis of the tank.

[30]"Tanks, Employment, Value of," 22 June 1915 communication of Field Marshall French to War Office, WO 158/831, PRO.

my command in France," and he asked for as many as could be supplied by May 1916 without prejudice to other war production.[31] This favorable attitude continued through the first use of tanks in combat, during the Battle of the Somme in September 1916. Employed in small numbers in muddy, heavily shelled terrain, the tanks bogged down. Of forty-nine tanks committed to battle, only nine managed to keep up with the infantry on the first day of operations. Yet Douglas Haig remained enthusiastic, writing to the War Office in October 1916: "From the experience of the employment of Tanks during the recent operations I have the honour to report that this new engine has proved itself to possess qualities which warrant further provision on a large scale." He supported the expansion of the Heavy Section of the Machine Gun Corps, as the Tank Corps was then called, "with a view toward the eventual provision of 1,000 Tanks, together with the necessary personnel."[32] The diary of one of the early proponents of the tank, Albert Stern, records the urgings of Haig: "He would do anything to help me; [he said] that a division of tanks was worth ten of infantry, and he probably underestimated it—told me to hurry up as many as I could—not to wait to perfect them but to keep sending imperfect ones as long as they came out in large quantities."[33]

While the tank was in development and small-scale production, it was essentially cost free, and there was no reason not to endorse it. Large-scale production, which meant reductions in resources for other missions, brought other responses. There were no mass tank armies in 1917 or 1918, and the British did not produce in 1917 as many tanks as had been proposed. Although a new measure of strategic effectiveness had been settled upon, there was no experience that demonstrated that the tank was in fact a more efficient way of using manpower until the Battle of Cambrai in the fall of 1917. Analysis of that battle that convincingly demonstrated that tanks and other motorized combat vehicles in fact made the most effective use of manpower on the battlefield was not available until March 1918. When such analysis was developed and presented, it was effective in obtaining more resources for tank production and tank unit, but until it was available, the additional resources required by tank production and deployment were seen as detracting from existing com-

---

[31]Notes by Haig dated 25 December 1915 and 9 February 1916, WO 158/831, PRO.
[32]Haig to War Office, 2 October 1916, WO 158/836, PRO. See also Charteris' endorsement of the tank after their use in the Somme in his diary entry for 16 September 1916, in Charteris, *At GHQ*, pp. 164–65, 168.
[33]Quoted in Prior, *'World Crisis' as History*, p. 245.

bat arms. The British did eventually learn that the tank was the best available weapon according to its measure of strategic effectiveness, but there first had to be convincing combat experience and analysis. By that time, however, the outcome of the war was already largely determined.

By late 1916 British losses had been extremely heavy, and the government instituted conscription. A law for the conscription of married as well as unmarried men had been passed in May 1916. Conscription was necessary not only to provide more men for the army but also to prevent workers needed for war production from joining the army. The ship-building and food production industries, under pressure from the German submarine war, were scrambling for more labor.[34] The Somme offensive was followed by a series of British offensives that were meant in part to pin the German army down so it could not exploit the weaknesses of the French army, which was suffering from the mutinies that followed the unsuccessful April-May Nivelle offensive after which the French had had to withdraw 250,000 men from the front.[35] Contemporary estimates were 900,000 British soldiers killed, wounded, or missing in the period from July 1916 to October 1917.[36] In March 1917 the Imperial General Staff noted in its "General Review of the Situation in All Theaters of War": "The most important question of the present time is manpower. . . . I regret to say that the efforts made to obtain men have proved inadequate . . . and if heavy fighting takes place in April or before, that the strength of the armies in the field diminish, no matter what steps are now taken."[37] By December 1917, Sir Henry Wilson, preparing to take over as chief of the Imperial General Staff, calculated that the British armies in the field would be 200,000 men below authorized strength by March 1918 and 400,000 men short by October 1918. Ten days after the 1918 German offensives began, Wilson realized that all available trained soldiers in England had been sent to France, and that only untrained eighteen-and-a-half-year-old boys were left as a reserve or for the defense of Great Britain.[38]

In this context, the reaction to the proposal for more tanks was that they would require too many men. The General Staff did en-

---

[34]Robertson, *Soldiers and Statesmen*, 1:293, 1:299, 1:301.

[35]John Williams, *Mutiny 1917* (London: Heinemann, 1962), pp. 155–56, 244.

[36]Charles Callwell, *Field Marshall Sir Henry Wilson*, 2 vols. (New York: Scribner's, 1927), 2:18.

[37]General Staff note of 20 March 1917, WO 106/1512, PRO.

[38]Callwell, *Sir Henry Wilson*, 2:41, 2:81.

dorse Haig's initial request for 1,000 tanks and five tank brigades on 20 September 1916, and the War Office had issued an order to the effect on 14 October 1916.[39] This decision, when finally implemented, gave each British army in France a coherent tank brigade with 144 tanks and 72 in reserve. Initial plans were to have 500 of those tanks produced by May 1917, although labor shortages kept actual production to less than half that. Until February 1917, Haig continued to make tanks one of his highest production priorities, preceded only by airplanes and previously authorized railroad equipment. In June 1917, he recommended that the Tank Corps be increased again to eighteen brigades, and the War Office authorized this expansion the same month. But by the beginning of July, the cumulative effects of the infantry losses of 1917, at the third Ypres offensive in particular, made it clear that personnel for any expansion of the Tank Corps would come not out of unmobilized manpower but from the infantry. There was nowhere else. It was only at this point that Haig and the War Office agreed to temporarily curtail the plans for tank production.[40]

If manpower was the crucial resource why did the advocates of the tank not make the argument that it made the most efficient use of this resource? The commander of the Tank Corps, Brigadier H. J. Elles did, in fact, make this argument. Machine power could substitute for animal power in moving artillery, with a net saving of perhaps 50 percent of the men now needed to care for horses.[41] Tanks could substitute for artillery. The massive artillery barrage before the third battle of Ypres had required 121,000 men. At the battle of Cambrai, along a front 40 percent as long as the one at Ypres, 4,000 men of the Tank Corps with their machines were able to attack without artillery preparation, clearing the path for infantry at a cost of 5,500 casualties, versus several hundred thousand lost in three months of fighting at Ypres.[42]

But this argument was not made until March 1918, after the November battle of Cambrai. Prior to this, tanks had not been used on

[39]Minutes of GHQ General Staff conference 19–20 September 1916, WO 158/836, PRO; "Tank Output Supply," 158/801, p. 4.

[40]"Organization of the Heavy Section of the Machine Gun Corps Later Known as the Tank Corps," WO 158/804, PRO, pp. 6–7, 11, 15–17. It is important to note that industrial and army demands for labor were not adjudicated by a central authority until August 1917, and that the army could be and was denied a request to have 215,000 men released from industry for the army. See Robertson, *Soldiers and Statesmen*, 1:300, 1:311, 1:314.

[41]"Tank Program 1919," Appendix C, "Petrol versus Muscle," WO 158/865, PRO.

[42]"Tank Program 1919," WO 158/865, PRO, p. 4.

a large enough scale and with the understanding that they did not require the same artillery barrage in preparation that infantry attacks did. Until that happened, there were no lessons to be learned that would show the efficiency of tanks when compared with conventional assaults. On the strength of this experience, Elles proposed a new tank army and a new concept for tank warfare. New, more mobile tanks would penetrate deeply and rapidly, supported by yet to be developed armored troop carriers. A force of 17,596 tanks of all kinds should be built for a blitzkrieg offensive to be launched in the spring of 1919.[43] The British alone could not produce such a force, but together with France and the United States Elles believed it could be done. He was able to convince the minister of munitions, Winston Churchill, to advocate a program to produce 4,000 tanks in Britain, and in March 1918, the chief of the Imperial General Staff, Henry Wilson, approved this plan on the strength of Elles' arguments concerning manpower.[44]

Developing a strategic measure of effectiveness and applying it to the question of the allocation of scarce resources was not the only problem of organizational learning related to the tank. At the tactical and operational level it was necessary to learn to use the tank properly, and this process also required an intellectual and organizational revolution. The British army was poorly organized for this lower level learning. Not only did it suffer from the dispersal of the General Staff at the beginning of the war, but it also resisted efforts to develop central organs that could use battlefield information and innovations from one sector to help all sectors improve performance. For example, no new army-wide infantry training manuals were issued by the GHQ of the British armies in France from 1914 to 1917. When Douglas Haig was made the commander in chief of the armies in the field, he abolished all corps-level schools in favor of division-level schools; the latter had so narrow a focus that it was impossible for them to make a theater-wide evaluation of ongoing operations.[45] When a crucial improvement in artillery tactics, the creeping barrage, which suppressed enemy fire ahead of an advancing line of infantry, was made by one corps level commander, its use remained confined to that corps for some time. The commander of the Fourth Army, General H. S. Rawlison, won approval to perma-

---

[43]Ibid., pp. 10–18. See also Elles's 3 January 1918 memo to the General Staff of GHQ, WO 158/835, PRO, in which these ideas are sketched out for the first time.

[44]Callwell, *Sir Henry Wilson*, 2:68; Albert G. Stern, *Tanks 1914–1918: The Log Book of a Pioneer* (London: Hodder and Stoughton, 1919), p. 212.

[45]Travers, *Killing Ground*, pp. 111–12.

nently assign an intelligence officer to support his artillery. The intelligence support made possible precise maps of enemy gun positions, and this, in turn, made possible the replacement of massive barrages, which took weeks to prepare (in order to accumulate the necessary shells) and days to execute, with rapid, accurate surprise barrages. Yet Rawlison's innovation also failed to touch the army as a whole.[46]

The British army was quick to use the tank as an adjunct to the infantry, to crush barbed wire and terrify the enemy. It was slower to learn the way in which "the internal logic of [tank] technology," as Timothy Travers has written, required not only the addition of tanks to existing forces, but also a transformation of the ways in which the older weapons were used. Without this transformation the tank could not be effectively integrated into the totality of military operations, and might suffer from friendly action. This learning and integration could not be carried out on a decentralized basis. The men of the Tank Corps understood as early as August 1916 how the other branches of the British Army had to change, but they were powerless to enforce their understanding. The artillery continued the practice of heavy artillery barrages before tank assaults, barrages the tanks did not need and which, in fact, only cratered and pulverized the ground so as to make it less passable for tanks.[47] In low-lying ground where the water table was close to the surface, heavy shelling churned the ground into a bog in which tanks could and did sink, as at Ypres 1917.[48] A service-wide publication on how to use artillery with tanks does not seem to have been issued until the General Staff manual titled "Tanks and their Employment in Cooperation with Other Arms" appeared in 1918.[49]

The problem of learning the proper tactical use of the tank was complicated by the fact that the Germans were an active enemy who developed countermeasures after the first use of the new technology. By 1918, they were using antitank rifles and artillery, which had to be suppressed by friendly artillery and infantry. A combined arms doctrine for the use of tanks was necessary, not just a concept of operations for the Tank Corps alone.[50] In order to develop and

---

[46]Bidwell and Graham, *Fire-Power*, pp. 83–85, 91, 104.

[47]General Staff Headquarters, August 1916, "Preliminary Notes on the Tactical Employment of Tanks," WO 158/834, PRO.

[48]J. F. C. Fuller, *Memoirs of an Unconventional Soldier* (London: Nicholson and Watson, 1936), pp. 131–36.

[49]WO 158/832, PRO.

[50]Fuller, *Memoirs*, pp. 313–14.

implement necessary concepts, new central institutions had to be created. J. F. C. Fuller, a staff officer with the Tank Corps at the time, sought to bring this about in 1918. He advocated the creation of a position on the Imperial General Staff with responsibility for tank warfare, along with a staff officer for tanks at the headquarters of the British armies in the field. The purpose was to create channels for handling information about the tanks similar to those already established for older weapons. This would make possible normal, organization-wide learning about tank warfare. The new organization, as Fuller put it, would place the new weapon "on the normal Army footing: the War Office would collect all tank ideas from all Tank Corps [sic] and formulate a tank policy, and GHQ would set this policy in motion after working out the details with its expert."[51]

At the tactical level as well as the strategic level, the British Army had to learn what and how to learn about the tank. By the end of the summer of 1918, Fuller's plan of organization was essentially in place, with Fuller himself in the Imperial General Staff office he had advocated. From this position he was able to learn about the emerging German countermeasures and how to defeat them. From his in depth study of the tank and the new German defenses in depth emerged his "Plan 1919," a scheme for the combined use of infantry, self-propelled artillery, and new, faster tanks with greater range to penetrate deep enough behind enemy lines to cut corps level command links.[52] This new concept would correct many of the weaknesses that had marred even the relatively successful use of tanks at Cambrai in November 1917. By 1918, the British army was in a position to learn how to use tanks against a reactive enemy and had developed all of the intellectual bases for blitzkrieg warfare.

CONCLUSION

Most analyses of the introduction of the tank during World War I tend to focus on the period in which the decision to produce the tank was made. What is most noteworthy in this regard is the speed with which the tank appeared. The idea was first raised in November 1914 by the man credited as the tank's inventor, Ernest Swinton. He received funding for the first prototypes in April 1915 and did

[51]Ibid., p. 285.
[52]Ibid., pp. 321–27.

not resolve upon a basic design until the end of June 1915. The first order for tanks for use in the field was placed by the minister of munitions in February 1916.[53] From the first conception to first production required only fifteen months. The technology was not resisted by senior military commanders. The problem was not in deciding to produce limited numbers but in learning how to evaluate the weapon's strategic merit relative to competing modes of warfare. A strategic measure of effectiveness, efficiency in using manpower in a strategy of attrition, had to be established before any lessons could be drawn from the first experiences of the tank. Then, the army had to reorient its operating concepts to make effective use of the tank. That learning and reorientation process took from August 1914 until early 1918, over forty months. A conception of how to use the tank at the tactical and operational level was not delineated until later in 1918. The delay in the successful implementation of this wartime innovation reflected not a failure to try the new technology but a failure, or rather a slowness, of organizational learning. This slowness, in turn, was directly related to the problems of defining a new strategic measure of effectiveness, of utilizing available information to evaluate the innovation, and the absence of tight central controls to ensure the implementation.

The focus on attrition and on conserving manpower meant that analyses of the use of the tank which focused on this aspect of its utility were very persuasive in supporting this innovation. That analysis of necessity followed rather than preceded the successful use of the tank in some numbers on the battlefield. Poor intelligence about German manpower tended to undercut the case for innovation. Better intelligence, indicating that the Germans were not on the brink of collapse in 1916, and a more rapid arrival at their new measure of effectiveness might have focused the British on the need for conserving manpower earlier, and thus to battlefield uses of the tank that exploited its ability to substitute for artillery and infantry.

In terms of the debate between Clausewitz and Sun Tzu on the possibility of wartime learning, the British case seems to favor Clausewitz. Intelligence about the enemy was vague and contradictory, and British understanding of their own strengths and weaknesses was slow to emerge. At the same time, this case suggest how wartime innovation might be facilitated. The intellectual and organizational changes necessary to evaluate new ways of fighting are as important as the development and production of new technologies.

---

[53]Swinton, *Eyewitness*, pp. 150–51, 215–17.

Early attention to these organizational questions could have facilitated an understanding of the relative value of the tank. Overall, its use of the tank showed that the British military was willing to learn from experience and was able to make innovations on the basis of lessons learned once the correct question had been formulated and there was relevant experience to be learned from.

Was the experience of the British army in World War I typical of the general problems of wartime innovation? A mass army fighting on a continuous front clearly needs some centralized means of coordinating its activities if it is to be effective. This need for coordination extends to the problem of innovation. Are there conditions, however, in which central coordination is less critical and in which innovation can proceed from the independent actions of unit commanders acting in response to their particular conditions? The next chapter on submarine warfare in the Pacific will explore that possibility.

[5]

New Blood for the Submarine Force

During the course of World War II, the American submarine fleet in the Pacific was transformed from a force targeted on the Japanese battlefleet into a force concerned with raiding merchant shipping. This innovation involved wholesale changes in the character of the fleet's officer corps and was effected in an organization with a radically decentralized command structure and good intelligence about the enemy. What impact did these factors have on innovation?

In World War II, innovation was necessary in the American military as a whole because prewar foreign policy had eschewed certain political goals and military capabilities. Though the technological capabilities necessary for certain innovations had emerged in the interwar period, concepts of operation that might have exploited the technologies did not emerge because they ran counter to basic principles of American foreign policy. Bomber aircraft and submarines, for example, were potentially offensive weapons that would give the United States the ability to intervene in a European war or to engage in unrestricted submarine warfare. However, the development of formal doctrines and institutional infrastructures in support of the strategic bombardment of European targets or the conduct of unrestricted submarine warfare was not politically possible in the United States in the 1930s. Thus at the beginning of World War II the nation was in possession of air forces equipped with heavy bombers and a navy with ocean-going submarines, but also with conceptions of operation very different from those that would ultimately prove most useful in war. Prewar doctrine for the U.S. Army Air Forces focused on the use of B-17 bombers for the defense of the Western

Hemisphere. Before Pearl Harbor, the U.S. submarine force planned for a war against the Japanese battlefleet. When war came, B-17s and other bombers were used not for hemispheric defense but to strike deep into Europe to cripple the industrial infrastructure of the Axis powers. U.S. submarines did fight the Japanese military fleet but also waged a campaign of unrestricted submarine warfare against Japanese merchant ships, *maru*, in order to strangle the Japanese economy. These fundamental changes in mission required wartime innovations. If the United States did not have to create new categories of weapons in wartime, it was nevertheless faced with the problem of rapidly creating new tasks and missions for existing forces. In effecting these innovations, the military had to learn from its wartime experiences.

The innovations were no less difficult for being confined to operating doctrine. During 1943, fully 30 percent of U.S. submarine commanders operating in the Pacific were relieved for cause because they were not able to make the transition from one mission to the other. Their peacetime training had made them unfit for the mission against merchant shipping. This innovation is thus linked primarily to changes in the character of the men in command of the technology.

This chapter and the next will also study these particular innovations to determine the impact of improved intelligence. The Allies enjoyed unusually good intelligence capabilities in World War II, and these cases suggest how innovation may proceed when intelligence support is perhaps as good as can be hoped for in wartime.

REMOVING THE PREWAR OFFICER CORPS

The use of submarines in the Pacific and Southwest Pacific commands of the United States Navy during World War II represents one of the clearest examples of a military organization being forced to fight engagements completely different from the ones it had trained and equipped itself for in peacetime. U.S. submarine commands before the war were strictly forbidden by international law and naval doctrine, as outlined in the "Instructions for the Navy of the United States Governing Maritime and Aerial Warfare," from engaging in unrestricted warfare against merchant shipping. As a result, "neither by training nor indoctrination were the submarines prepared to wage unrestricted warfare" when they were ordered to do so by the chief of Naval Operations on 7 December 1941.[1] Despite a hint that November from an officer of the navy's war planning staff to the commander in chief of the Asiatic Fleet (later to be subsumed in the Southwest Pacific Command) that unrestricted warfare would be permitted in the event of war,[2] the outbreak of war found the submarine force entirely focused on operations against Japanese warships. The name given to the submarines that were the mainstay of the force, Fleet Submarines, reflected their function in support of the U.S. Fleet in its efforts to engage the Imperial Japanese Navy.

Taking a force designed and trained for one mission and transforming it into one capable of performing another represents a special case of wartime innovation, one that does not involve new hardware. No new type of submarine was introduced into active service during the war, and while much effort was made to correct serious flaws in torpedo design, no new weapon with significantly greater capabilities was introduced.[3] The innovation was thus purely in the area of concepts of operation.

This innovation was shaped by the special character of submarine operations. Control of submarines on patrol is and must be decentralized, with great discretion given to submarine commanders. By design, submarines operate unobserved, by friends as well as enemies, and submarine commanders have, while on patrol, perhaps more autonomy than any other operational military commander. In the words of one historian, "a submarine skipper, far from direct supervision, in absolute command of his ship (and usually manning the periscope during an attack [*and so in sole possession of all the facts about the enemy's actions*]) could be as brave or as cautious as he wished and could fudge patrol reports to cover his actions."[4] Formal

---

[1]Commander, Submarine Force, U.S. Pacific Fleet, *Submarine Operational History, World War II*, mimeo, 2 vols., 1946 (hereinafter *Operational History*), available at Records Section, Naval War College Library, Newport, R.I. 1:1–2.

[2]Commander, Submarine Force, Atlantic Fleet, submitted to the Director of Naval History, *U.S. Naval Administration in World War II, Submarine Commands*, mimeo, 2 vols., 14 February 1946 (hereinafter *Administrative History*), Administrative History of the Navy in World War II series, Washington Naval Yard, Washington, D.C., 1:114.

[3]The basic source for most postwar accounts of the trials of the American submarine force as it coped with torpedoes that ran too deep, did not explode, or otherwise failed to perform in the intended manner is the *Operational History*, 2:696–750. The story of torpedo failures and the navy's response has been told often and well (see, for example, Clay Blair, *Silent Victory* [New York: Bantam, 1975], pp. 273–81, 435–39) and falls very much, as defined here, into category of "reform" as opposed to "innovation," and for that reason will not be discussed in this section.

[4]Blair, *Silent Victory*, p. 199; see also *Administrative History*, 1:196.

directives to implement innovative doctrine were more difficult to enforce than in other military operations, since it was difficult to determine whether orders had actually been carried out.[5] Under these circumstances, innovation in operational behavior could only be accomplished by effecting changes in the character of the submarine commanders themselves.

However, strategic decisions relevant to the submarine force are made on a centralized basis, just as for any other component of the navy. Determinations of the level of resources to be given the force, the relative emphasis to be given submarine operations in war, and the strategic evaluation of those operations relative to other navy and other friendly military operations have to be made on a centralized basis. There was thus a combination of radically decentralized operational control combined with the normal centralized policy-making structure for strategy. Early on, the navy had decided to use its submarines, aided by Ultra intelligence, the code name for American and British decrypting of high level German and Japanese military radio communications, to help sink Japanese battleships and aircraft carriers.[6] Was this the most strategically valuable use of submarines or, was it a waste of their potential ability to destroy the enemy's war economy by sinking its merchant ships? The answer depended on an understanding of the Japanese economy and a determination of the impact of unrestricted submarine warfare on the Japanese merchant fleet and its subsequent effect on the Japanese war-making capacity. The innovation thus operated at two levels. The first, the operational level, involved making individual submarine commanders capable of fighting the new kind of war; this took place in a decentralized administrative environment. The second, the strategic level, involved determining the extent to which this new capability would be used; this decision that had to be made in a centralized context.

How much change from prewar concepts was required by this innovation? The navy's prewar plans for the use of submarines in war are noteworthy for their sketchiness. One navy submarine designer,

---

[5]It has been reported, for example, that when faced with magnetic torpedo detonators that clearly did not work in combat and with standing orders not to deactivate them in favor of simpler contact detonators, many submarine commanders went ahead and deactivated them anyway, then doctored the patrol reports and swore the crews to secrecy. See Blair, *Silent Victory*, p. 206

[6]See, for example, the account of the pursuit of a Japanese task force including three aircraft carriers in April 1943 in W.J. Holmes, *Double-Edged Secrets: U.S. Naval Intelligence Operations in the Pacific During World War II* (Annapolis: Naval Institute Press, 1979), p. 134.

searching for clues about the ultimate use of the Fleet Submarines that would help him in his design work, wrote that aside from the obvious need for the submarines to be able to cruise with the fleet at all times, "as far as I know, we in our Navy have not as yet developed any concrete plan for the use of such submarines with the fleet."[7] The designers nonetheless produced submarines that proved successful in wartime. These designs did not result from an effort to focus on the specific character of the enemy and to build a submarine tailored to handle him.[8] Instead, the designers emphasized the special characteristics of submarine warfare against all conceivable enemies. The dominant characteristic of submarines, it was argued, was their ability to "penetrate sea areas not controlled by our Navy and even sea areas controlled by the enemy's navy." Using stealth to penetrate enemy waters, the submarine could perform reconnaissance against the enemy fleet and conduct attacks upon it. These operations "have an influence upon morale and material that cannot be measured but is of enormous effect," because the mere "presence of the submarine in the theater of operations produces a terrific strain upon enemy personnel and material." The special ability of submarines to enter enemy waters meant that they would be operating ahead of the American fleet and far from secure bases, and their ability to induce strain in the enemy by their presence meant that they would benefit from the ability to remain on station in those waters for as long as possible. Hence, there were reasons independent of the character of any specific enemy for building submarines with long cruising ranges (18,000 miles was suggested) and with habitability for the crew, which would have to live for weeks in the vessel without relief.[9] These abilities were designed into American submarines, and the resulting fleet was capable of long-range operations against either Japanese merchant shipping or the Japanese battlefleet.

[7]Commander Thomas Withers, "Design and Operations of Submarines," General Board Serial 1365, 23 May 1928, microfilm of the records of the General Board, Historical Collection, Naval War College, Newport, R.I. Withers became the commander of the Submarine Forces, Pacific Fleet, only to be relieved a few months after the attack on Pearl Harbor.

[8]In response to a General Board query, the president of the Naval War College urged the development of submarines to handle a war with Great Britain, as well as a war with Japan, since war with either could arise from trade disputes. Harris Laning to the General Board, "Design of Submarines," 23 September 1930, microfilm of the records of the General Board, Historical Collection, Naval War College.

[9]Commander Thomas Withers to the Secretary of the Navy, "Design of Submarines," 14 August 1930, microfilm of the records of the General Board, Historical Collection, Naval War College; Withers, "Design and Operations."

But if the submarine technology needed for long-distance operations against merchant shipping proved interchangeable with that needed for operations against enemy men-of-war, the same was not true for submarine personnel or their training. Submarines operating against a battlefleet faced extremely hazardous conditions. Heavily armed and armored, accompanied by aircraft that could spot a periscope or a surfaced submarine, and capable of moving at high speeds, a battlefleet had to be approached by submarines with great caution. Before the war it was thought that to avoid destruction submarines would have to utilize to the maximum their ability to stay hidden. Training exercises emphasized this point. In the 1940–1941 Gunnery Exercises, for example, it was estimated that a submarine that stayed completely submerged as it approached the enemy fleet, using sonar to direct itself, had only a 50 percent chance of approaching within attack range without being detected and a 25 percent chance of making its attack and escaping undetected. Once detected, the sub had an estimated chance of evading a depth charge attack of only one-in-seven.

Anything that made detection more likely had to be avoided. In particular, submarine commanders learned that the use of periscopes to search for and attack enemy ships would be suicidal. Exercise reports noted, for example: "The Division Commander considers that an undiscovered attack at periscope depth was virtually impossible." "Commander Submarine Division considers that due to increased air protection that is being given Fleets . . . , it is practically mandatory for the attacking submarine to fire on sound [*i.e.* sonar] information alone." And stern warnings were issued to those who deviated from the received wisdom: "Commander Submarine Division is directed to take the steps necessary to assure that commanding officers of vessels under his command will, on future practices with similar conditions existing, more properly conduct their approach, and if practicable fire by sound."[10]

The extreme caution built into prewar concepts of operation was artificially reinforced by unintended factors in the training process. The open-water training for Pacific submarines was conducted in waters completely different from those in which the war would actually be fought. The training waters lacked, for example, any of the thermal layers that would hide submarines from even vigilant destroyers equipped with sonar. Peacetime training exaggerated the ability of aircraft to detect submarines. One submarine commander

[10]*Operational History*, 2:585–86.

described the way in which fear of being sighted by hostile aircraft was artificially trained into prospective commanders:

> We were being held back by peacetime training against warships that had surface and plane escorts. The planes [in training] were from the utility squadron . . . and did this screening on a regular basis. The pilots could observe both the target group and diving submarines at the start of the run, and knew the firing positions [of the submarine] within a mile or so, and the firing time within a couple of minutes. When the sea was without whitecaps, they would dip a wing and point out the submarines to us—the junior officers from other boats who were observers. So having periscopes sighted by airplanes became a bugaboo for some, while, in truth, the scopes would never have been sighted on the open seas.[11]

In wartime, these cautious tactics proved to be counterproductive and had to be abandoned. Despite the emphasis placed on the technique in peacetime training, only 31 out of 4873 known submarine attacks were directed by sonar. The habits of mind produced by training were important. In the words of the U.S. Navy's internal history. "The false lessons learned in peace undoubtedly influenced the submarine to use extreme caution in the very early patrols."[12] The caution, in fact, persisted beyond those early patrols and constituted a significant obstacle to the use of the submarines as a strategically valuable weapon.

Why was it that submarine commanders trained in the use of cautious, stealthy tactics for attacks on fast moving, heavily armed and protected warships did not simply have a field day attacking slow, unarmed or lightly armed merchant ships? The difficulty arose from the fact that in the vast expanses of the Pacific, the individual merchant ships or small convoys employed by Japan were difficult to locate. To solve this problem, attempts were made in 1942 to operate submarines in waters closer to the Japanese home islands, in the hopes that, although more dangerous, those waters would be more densely populated with merchant targets. These deployments, however, proved not to be particularly lucrative. But the capture late in 1942 of Japanese charts depicting merchant shipping routes in-

---

[11]Richard H. O'Kane, *Wahoo: The Patrols of America's Most Famous World War II Submarine* (Novato, Calif. Presidio Press, 1987), pp. 29–30, 53. O'Kane served as the executive officer on the *Wahoo* and went on to become the submarine skipper with the most kills to his credit, the commander of the *Tang*, and recipient of the Medal of Honor.

[12]*Operational History*, 2:587–88.

creased the number of sightings,[13] and early in 1943 the cipher used for radio transmissions to and from Japanese merchant ships was broken by the U.S. Fleet Radio Unit Pacific (FRUPAC), the intelligence unit based in Hawaii.[14] These messages routinely listed the names of the merchant ships, their cargoes, their routes, and their expected noontime positions on specified dates.[15]

Even with this intelligence support, however, submarine commanders using prewar concepts of operations had serious problems finding merchant ships. Moving slowly underwater, using sonar or their periscope only minimally, submarine commanders simply could not see or hear very far. Even when directed to the sector in which a merchant ship was expected, the subs could not effectively search the area before the ship had passed. Merchant ships moved more quickly than submarines moving quietly underwater, and if the submarine missed its intercept, it could not stay submerged and hope to trail its target. Even if the submarine did make contact with a target, if the commander adhered to prewar tactics and stayed submerged and invisible, he was likely to lose it. One submarine officer described a typical encounter between a merchant ship and a submarine conducted by a commander trained in the old school. While surfaced, the commander sighted a merchant vessel at the edge of the horizon and immediately submerged until only three feet of periscope was showing above water. "Of course he saw nothing with three feet of scope. I suggested a higher search, but he chose to wait for the enemy to close." The target then got away. On another occasion, the same commander made contact with the merchant ship within sight of a deserted jungle island. A chase by the submarine on the surface, where its speed was higher, was suggested to the commander. " 'Why, they'd have planes over us in minutes' he

[13]*Operational History*, 1:17, 1:31–32

[14]Holmes, *Double-Edged Secrets*, p. 126

[15]Ronald Lewin, *The American Magic: Codes, Ciphers, and the Defeat of Japan* (New York: Farrar, Straus and Giroux, 1982), p.224. The relevant extracts from the decrypts of intercepted Japanese radio communications are available in the Naval War College Library and in SRH-011, *The Role of Communications Intelligence in Submarine Warfare in the Pacific*, 8 vols., Record Group 457, National Archives, Washington, D.C. This document, however, is simply a compilation of the reports of expected times and positions and number of Japanese ships, and contains no analysis or assessments. For one terse assessment of the high value of this communications intelligence for the conduct of the U.S. submarine war, see "ComInt Contributions, Submarine Warfare in World War II," 17 June 1947, Vice Admiral Charles A. Lockwood to Chief of Naval Communications, SRH-235, Record Group 457, National Archives, Washington, D.C., reprinted in Ronald Spector, *Listening to the Enemy: Key Documents on the Role of Communications Intelligence in the War with Japan* (Wilmington, Del.: Scholarly Resources, 1988), pp. 133–35.

scoffed." The target escaped again.[16] These incidents dated from late 1942 and early of 1943. Similar behavior was displayed by other commanders at the time, despite the fact that wartime experience had by then established that "the danger of being picked up at periscope depth by patrolling aircraft had been overdrawn."[17]

Effective tactics for searching larger areas of ocean and for turning contacts into successful attacks had been developed by a few submarine commanders as early as March 1942, tactics that were wildly dangerous by prewar standards: "Having missed . . . its target, or sunk only part of the convoy, the submarine could wait until [the target had moved and] it was out of sight, surface in broad daylight, make a high speed, 'end around,' and attack again . . . a submarine could steal silently into a convoy on the surface at night, sink two or three ships, and escape at high speed without even diving, and perhaps repeat the performance a few hours later on the convoy remnants."[18] But individual learning and organizational learning are two different things, and the operational behavior of the majority of U.S. submarine commanders continued to conform to prewar concepts. Organizational failure did not lead immediately to organizational learning and innovation. In another case of feedback producing a vicious circle, the harder submarine commanders tried to live up to established criteria for operational conduct, the worse their problem became.

Why did the failure to find and sink enemy ships not produce an order from the center to employ a new doctrine? To begin with, the highly decentralized character of attack submarine operations precluded the issuing of any detailed doctrine. The commander of Submarine Forces Pacific, for example, never went further in formal directives than to state that the goal of submarine operations in his area was to "inflict maximum damage to enemy ships and shipping by offensive patrols at focal points," to lay mines, and to perform other special task as required.[19]

There were mechanisms for passing on lessons learned that stopped short of trying to impose a new operating doctrine, however, each submarine commander upon his return from patrol would submit a

[16]O'Kane, *Wahoo*, pp. 54, 100, 191. The importance of the "higher search" was that by rising to show seventeen feet of periscope, instead of three feet, the submarine would treble the distance it could see, from five to fifteen miles.

[17]*Operational History*, 2:518.

[18]*Operational History*, 1:4, 1:19–20, 1:124–27; 2:641.

[19]*Administrative History*, 1:45. The 1943 Casablanca Conference issued similarly vague directives on submarine warfare, calling only for continued pressure on Japanese sea lines. See Blair, *Silent Victory*, p. 399.

report to the commander of submarine forces in his area. In the Pacific Command, the strategic planning officer working for COMSUBPAC would select portions of those reports for publication in the Tactical Bulletin.[20] The patrol reports themselves would be returned with an "endorsement," an assessment of performance by the commander's superior officers, and these endorsements became an instrument for transmitting new policies from the top down.[21]

What is noteworthy is how ineffective the Tactical Bulletin and the system of endorsements was in changing the behavior of individual submarine commanders. The commander of the very first submarine to go on a war patrol from Pearl Harbor after the Japanese attack was severely chastised in his endorsements for having been too cautious about aerial attack and spending too much time unproductively, hiding underwater.[22] Endorsements from that time on consistently urged more aggressive submarine tactics in the harshest of language.[23] Yet, as we have seen, individual submarine commanders operated with extreme caution into 1943. The problem, it was generally acknowledged, was that the autonomy given to submarine commanders made it useless to order them to change tactics when they did not have the necessary emotional stability and daring in their characters. Moreover, it proved to be impossible to determine ahead of time which men would display the necessary character traits in combat. Although the navy tried many methods to identify the right kind of men for submarine commands, no satisfactory technique was ever found.[24]

In order to translate the emerging understanding of the new way of fighting into a change in the behavior in the submarine officer corps, a change in the composition of that corps was necessary. As one submarine commander put it, whenever a submarine had changed its behavior, "it had come about with a change of command."[25] In war, demographic change in the composition of an officer corps is not unusual, but it is important to distinguish at which level the changes are taking place. The pattern in both world wars was for significant turnover among the officers with field commands, but relative stability in the high command. Because operational doctrine is usually made by the high command, turnover in

[20]*Administrative History*, 1:75, 1:77.
[21]Blair, *Silent Victory*, p. 119.
[22]*Operational History*, 2:588–89.
[23]Blair, *Silent Victory*, p. 120.
[24]*Administrative History*, 1:168–73, 1:195–97.
[25]O'Kane, *Wahoo*, p. 81.

the officer corps in war need not be associated with changes in operational behavior. In the French army in World War I, roughly 10 percent of the officer corps was killed outright in August 1914.[26] The commander in chief of the French army, Marshall Joffre, recounts in his memoirs that in the first weeks of the war he was obliged to relieve two army-level commanders, nine out of twenty-one corps commanders, thirty-three out of seventy-two infantry division commanders, and five out of ten cavalry division commanders. But it was Joffre who made army doctrine, and he made it clear that he relieved these officers because of their inability to perform up to standards Joffre had developed in peacetime.[27] This drastic purge of the French army thus did not produce any change in doctrine. Joffre himself stayed in command until 1917, when his hand-picked successor, General Nivelle, took over and, employing essentially the same concepts of operation, launched the offensives named after him that broke the French army as an instrument of offensive warfare.

In the case of the British army in World War I, approximately the same pattern prevailed, with rapid turnover at the levels below the high command and stability at the top. One British general estimated, for example, that in the infantry brigade with which he was familiar, three-quarters of all battalion-level commanding officers were weeded out as unfit for command by September 1914.[28] Yet there were only two commanders of the British armies in the field, John French and Douglas Haig. While John French was relieved in 1915, it was less over his conduct of the war and more over an operation decision that had resulted in severe tensions with the French government.[29] Lord Kitchener remained secretary of state for war for over two years. A good indication of the stability at the top of the British military high command can be had by looking at the eight senior military officers the British cabinet assembled to advise it on the eve of war on August 5, 1914. Two would die or retire shortly

[26]Cited in Robert Asprey, *The First Battle of the Marne* (Philadelphia: J.B. Lippincott, 1962), pp. 58–59.

[27]Marshall Joffre, *The Personal Memoirs of Joffre*, 2 vols., trans. T.B. Mott (New York: Harper, 1932), 1:29–30, 1:276. On the Dreyfus scandal and the origins of the need to weed out the officer corps drastically, see Douglas Porch, *The March to the Marne: The French Army 1871–1914* (Cambridge: Cambridge University Press, 1981), pp. 214–15.

[28]Timothy Travers, *The Killing Ground: The British Army, the Western Front, and the Emergence of Modern Warfare 1900-1918* (Boston: Allen and Unwin, 1987), p. 14.

[29]There was some unhappiness with John French's heavy consumption of artillery shells, but the root of his dismissal was political. See, Michael and Elanor Brock, eds., *H.H. Asquith: Letters to Venetia Stanley* (Oxford: Oxford University Press, 1985), p. 488; David French, *British Strategy and War Aims, 1914–1918* (Boston: Allen and Unwin, 1986), pp. 111–12, 162–63.

after the beginning of the war. The rest, Field Marshall Sir Henry Wilson, Lord Kitchener, Sir John French and his two corps commanders, Field Marshall Douglas Haig and General Grierson, and Ian Hamilton, would play prominent roles throughout the war. The only major figure not present at the meeting, Field Marshall William Robertson, began the war as quarter-master general to Douglas Haig and served as the effective chief of the Imperial General Staff from the end of 1915 until February 1918 when he was replaced by Henry Wilson.[30]

In the case of the U.S. Army Air Forces in Europe in World War II, bombing raids that suffered average losses of 5.5 to 6.5 percent per sortie over Europe, and over 10 percent over Germany,[31] inevitably generated enormous turnover in the officer corps, even at the level of flag-rank officers, since brigadier generals often flew with their units. But at the level of high command, what is remarkable is the stability among the commanders of the strategic bombing force deployed in Europe, generals Carl Spaatz, Fred Anderson, James Doolitle, and Ira Eaker, though they did circulate among various commands.

The officer corps of the U.S. submarine forces in the Pacific exhibited the same split pattern of demographic change. At the more senior levels there were some initial changes and then a prolonged period of stability, with some rotation of jobs among a small circle of officers.[32]

What was unique about the submarine force was that the autonomy of command at the lower levels meant that turnover at that level did lead directly to fundamental operational changes. Turnover was generated through a decision by senior submarine force commanders to give each skipper no more than two war patrols to prove himself by sinking ships. If he did not produce results, he was relieved. Many skippers did not wait, but asked to be relieved, or they suffered emotional collapse as a result of the pressure of combat.

---

[30]William Robertson, *Soldiers and Statesmen 1914-1918* (New York: Scribners, 1926), 1:53, 1:151, 1:187–88, 1:232–37.

[31]Stephen L. McFarland, "Evolution of American Strategic Fighter in Europe 1942–1944," *Journal of Strategic Studies* 10 (June 1987), 190,192.

[32]For example, in COMSUBPAC, Thomas Withers was relieved in the first months of the war, largely because he was associated with the unsuccessful prewar training, and replaced by Robert English. He died in an airplane accident in January 1943 and his replacement, Charles Lockwood, served out the war in that post. Lockwood himself came from a senior submarine command in the Southwest Pacific, and his departure led to some rotation of jobs involving Ralph Christie and James Fife, who continued to serve in that area for the rest of the war. See Blair, *Silent Victory*, pp. 223, 365–367.

The results were striking. Thirty percent of serving submarine commanders were relieved for cause in 1942, the year of greatest adjustment. Fourteen percent were relieved in 1943 and the same in 1944.[33] They were replaced by ever younger officers, who, as it turned out, were more likely to display character traits necessary for the new kind of war. From December 1941 through March 1942, forty-eight submarine commanders went on their first war patrol. Their average year of graduation from Annapolis was 1926. Fourteen months later, the new skippers leaving Pearl Harbor were, on average, four years younger.[34] The most productive submarine skippers were even younger. The top seventy-seven submarine commanders in the Pacific (out of a total of 465) had five or more confirmed kills each, and collectively they accounted for 51 percent of all enemy ships sunk by U.S. submarines. This group had an average Annapolis graduation year of 1931. The top 10 percent of all submarine commanders, measured by number of confirmed kills, had an average year of graduation of 1932 and accounted for 32 percent of Japanese ships sunk by U.S. submarines, with an average of ten kills apiece.[35]

A sharp shift in operational procedures was produced through a personnel selection process that rewarded success in combat, rather than through centrally developed and promulgated doctrine. The selection process was not based on a new strategic measure of effectiveness. There was apparently no attempt in 1942 to redefine the strategic goal of the submarine force away from defeat of the enemy battlefleet to the destruction of the Japanese war economy, to relate submarine operations to that goal, and to evaluate alternative methods of reaching that goal. Submarine skippers were simply told to sink ships, any kind of ships, or they would be relieved of command. This selection process produced a demographic change in the officer corps as younger officers tended to survive the selection process. This yielded an increase in merchant ship sinkings. Innovation was decentralized. Younger officers made a new kind of war on their own.

[33]Ibid., p. 533.

[34]My calculations, based on Joint Army Navy Assessment Committee (JANAC) tables reprinted as Appendix F in Blair, *Silent Victory*, pp. 901–7.

[35]My calculations from JANAC tables reprinted as Appendix G in Blair, *Silent Victory*, pp. 984–87. The pattern of a small portion of active fighters inflicting a disproportionate amount of damage on the enemy is also found in another combat arm in which individual weapons systems are controlled by officers enjoying a high degree of operational autonomy—fighter pilots flying air superiority missions. I am indebted to Barry Watts for this observation.

But there was a price to this decentralized innovation. Because it was not the result of a decision by the high command to explicitly redefine the strategic goals of the submarine force or an analytical effort to evaluate the impact of alternative military measures on the Japanese war economy, the impact of the innovation was missed until late in the war. Submarine commanders were left on their own to produce results or get out of the service. They responded by sinking more merchant ships. But in the absence of a redefinition of the strategic measure of effectiveness for submarines, the rest of the navy did not shift to focus on the strategic importance of what the force was accomplishing.

To be sure, there was some sense that at the operational level the submarines were winning a great victory. The sense of operational success was even greater during the war than after the war. Wartime claims of ships sunk by submarines were exaggerated, and had to be reduced from 10.17 million tons of Japanese shipping to 4.8 million tons of merchant shipping and 0.5 million tons of warships by the postwar Joint Army Navy Assessment Committee.[36]

But merchant ships were the intermediate, operational goal. Merchant ships had value because they held together the war economy of the Japanese empire. The strategic goal was to destroy the ability of the Japanese economy to support the Japanese military effort. An understanding of the impact of the innovation relative to the other components of U.S. strategy in the Pacific lagged behind performance, and the discrepancy may have resulted in a failure to capitalize on the strategic success of the innovation. Postwar data revealed, for example, that the sinking of Japanese oil tankers had by 1944 forced the Japanese government to devote over five-sixths of the steel available for shipbuilding to the construction of merchant vessels, despite the heavy losses suffered by the Japanese navy at Midway and the Battle of the Philippine Sea. The combination of submarine attack and strategic bombing had reduced the stockpiles of oil in Japan from forty-three million barrels at the end of 1944 to less than four million barrels in March 1945.[37] By the summer of 1945, there were virtually no Japanese ships left of over one thousand tons. Without merchant shipping, food, oil, and munitions could not be transported within the Japanese empire. Postwar data

[36]Ibid., p. 900.
[37]Lewin, *American Magic*, pp. 224, 229.

suggests that the campaign against the Japanese economy, of which the submarine force was an important part, had come close to breaking the empire, without invasion of the home islands, without Soviet intervention, and without use of the atomic bomb.

The analytical capability to understand the strategic impact of the innovation in submarine warfare was not in place until almost the end of the war. The intelligence task was formidable. Accurate data were needed about each merchant ship destroyed, including its identity. Without knowing which ship had been sunk at what time and place, it was impossible to assess the cargo destroyed, the carrying capacity destroyed, or whether the ship had already been credited as a kill to another sub or airplane (to guard against double counting). It was also necessary to know the size of the prewar merchant and the rate of construction of new vessels. Finally, stockpiles and rates of consumption of raw materials carried by merchant ships would have to be determined, so that a calculation could be made of the rate at which resources would be depleted by raids on merchant shipping.

Good data about the size and number of Japanese merchant ships were easily available to U.S. intelligence after the Japanese merchant shipping code was broken. In the absence of this type of information, close tracking of other data, such as records retrieved from sunken ships, observations made by the submarines themselves, and the monitoring of uncoded communications could, by cross-referencing, yield reasonably accurate accounts of ships sunk. This cross-referencing was time consuming and required complete, IBM card-readable files.

Unfortunately, while the branch of the Combat Intelligence Center at Pearl Harbor that was tracking Japanese merchant ships had access to Ultra cryptographic intelligence, it was short of manpower and could not maintain a cumulative account of the impact of submarine operations on the net capacity of Japanese merchant fleet. The statistical branch of the Navy Staff, Op-16-P, which compiled cumulative data on Japanese merchant ship losses, was not fully cleared for Ultra intelligence. The branch of the office of the commander in chief of the U.S. Fleet (Admiral King's office) that handled combat intelligence and that was fully cleared for Ultra intelligence was, until November 1944, devoted to the antisubmarine war in the Atlantic. The right combination of manpower, security clearance, and analytical focus was not brought together until November 1944.[38] At that time, a special office, the Pacific Strategic In-

---

[38]The account of the methods used to analyze merchant ship sinkings and the bureaucratic delays involved in the proper utilization of Ultra intelligence is laid out

telligence Section, was established under James Rochefort, the man who headed the team that provided the decrypted intelligence for the U.S. fleet at the Battle of Midway.[39] A Navy officer described the inadequacies up to this point: "Facilities had been available for the valuation and interpretation of only immediate operational intelligence which comes from these ultra [sic] sources. Much valuable information has been stored away . . . for the day when personnel could be made available to commence strategic studies based on information from this source. Under the present system, full use and evaluation of this material has not been obtained."[40]

Good tactical use had been made of the Ultra decrypts to alert submarines to the targets within their range, but a strategic assessment based on a cumulative analysis of the decrypted traffic had been deferred by the agency with full Ultra clearance because it was understandably preoccupied with using Ultra intelligence to help fight the German U-boats in the Atlantic. Even the commander of Submarine Forces Pacific was slow to start a group of analysts working on a quantitative analysis of his own submarine war. The Submarine Operations Research Group was not set up at Pearl Harbor until November 1943. It then began the time-consuming job of catching up with two years of accumulated data, and it had not entered all its data onto IBM cards by the end of the war.[41]

Thus the strategic impact of the submarine campaign in the Pacific was not assessed until the war was over, due to a lack of emphasis, problems with the circulation of classified data, and manpower shortages. An important innovation had taken place in the field but had overtaken the ability of policy makers at the center to assess its importance.

Even if an adequate effort to study the cumulative sinking of Japanese merchant ships had been made, it would still have been necessary to understand Japanese shipbuilding and the relationship of merchant shipping to the Japanese economy. But the U.S. government knew very little about the rate of merchant-ship construction during the war. How much had it accelerated relative to peacetime? We knew that U.S. construction had been speeded up significantly

in SRMN-039, "COMINCH Pacific Strategic Intelligence Section (PSIS) File," March 1944–December 1945 (hereinafter SRMN-039), Record Group 457, National Archives, Washington, D.C., 14 November 1944 COMINCH to Vice Chief of Naval Operations establishing the PSIS, see pp. 001–008, 028–030.

[39]Edwin P. Layton, with Roger Pineau and John Costello, *"And I Was There": Pearl Harbor and Midway—Breaking the Secrets* (New York: William Morrow, 1985), p. 468.

[40]SRMN-039, W. R. Smedberg, 20 November 1944 to F-2 (head of combat Intelligence in the office of COMINCH), p. 033.

[41]*Administrative History*, 2:413–16.

through the use of prefabricated sections welded together in the shipyards. Were the Japanese doing this? Was labor available to keep the yards working on a round-the-clock basis? American intelligence did not know.[42]

What was the impact of destruction of Japanese oil tankers? The answer depended on the size of Japanese oil reserves. The only base figure American intelligence had was from 1929. This figure was adjusted for the interim period with available import, production, and consumption figures. Unfortunately, in the words of one of the analysts involved, "the final figure arrived at was 50% higher than any previously used in either Washington or London. Obviously, such a method of calculating stocks would result in a strong possibility of error."[43] As for wartime oil production, Japanese production from the Netherlands East Indies was entirely unknown, and synthetic petroleum production could only be estimated.[44] Under the circumstances, it was perhaps not surprising that after Japan had surrendered, it was discovered that its stocks of aviation gasoline were only half of what had been estimated in June 1945, and that production of aviation was only 60 percent of what had been estimated during the war.[45]

CONCLUSIONS

The process of wartime innovation observed in the case of the tank and the British army in World War I highlighted the importance of developing new strategic measures of effectiveness, because of the need to allocate scarce resources among old and new ways of war. In the case of the American submarine war in the Pacific in World War II, the innovation took place without any such effort. Strategic wealth enabled the United States to pursue the war against Japan by means of the advance across the Pacific and the Southwest Pacific, by means of strategic bombardment and mines laid from the air, and by raiding Japanese maritime commerce with the submarine force. The innovation in the submarine force was thus never ham-

---

[42]Guido Perera, *History of the Organization and Operation of the Committee of Operations Analysts*, USAAF, 1945 (hereinafter *History of COA*), available on microfilm from the Air Force Historical Division, Bolling AFB, Va., pp. 69–76; see chapter 6 below.

[43]Ibid., p. 77.

[44]Ibid., p. 78.

[45]SRMN-039, compare 20 June 1945, PSIS to F-22, p. 120, with 23 November 1945, L.H. Frost to Smedberg, p. 143.

pered by the failure to set priorities among alternative strategic paths. At the same time, decisions were made to invade Japan, to bring the Soviet Union into the war, and to use the atomic bomb that might have been made differently if the effectiveness of the war against the Japanese economy had been correctly assessed. Given wartime uncertainties and pressures, those decisions can easily be justified, but they did have costs. This case shows that assessing the relative strategic effectiveness of an innovation can be important not only for the promotion of the innovation, but for the proper formulation of larger strategy.

[6]

The United States Strategic
Bombing Force, 1941–1945

As were chapters 4 and 5, this chapter is intended to explore how the problem of redefining strategic measures of effectiveness helps us understand wartime military innovation. If the presence or absence of good intelligence and the presence or absence of centralized command structures are thought of as structural conditions within which organizations try to develop new measures of strategic effectiveness, the cases involving innovation in the United States Army Air Forces in World War II take place in the context of relatively good military intelligence about enemy military activities, better intelligence than the British army had in World War I, but of poor intelligence about the enemy economy. The ability to intercept and decrypt enemy radio communications, most spectacularly in the case of Ultra intelligence, provided the American and British military services, and in particular the Army Air Forces, with intelligence about enemy operations and capabilities that was far superior to the intelligence available to the British about German casualties in World War I, but this intelligence was not always found relevant for the purposes of innovation. The innovations in the Air Forces also took place in the context of a force in which intelligence analysis was decentralized into several competing groups, but in which doctrine and operating concepts were centrally developed and directed in wartime, first by the Eighth Air Force and then by the United States Strategic Air Forces. In comparison, the innovation in the British army took place in the context of poor intelligence and decentralized command structures, and the innovation in the United States Navy submarine forces took place in the context of good intelligence, poorly distributed, and radically decentralized command.

The U.S. Army Air Forces strategic bombing forces developed two new military capacities in World War II: the ability to provide fighter escorts to bombers on mission and the ability to analyze the enemy to determine the targets the destruction of which would present him with the greatest difficulty in waging war. Target analysis is the less obvious innovation, because it involves no new weapons or military hardware. Yet it is no less a military innovation than the creation of a radar air defense network or a fire-control system for antiaircraft artillery. Each has the function of locating and evaluating enemy targets and directing friendly weapons against them, though one uses electronic signals and the other intelligence data and economic analysis. Each has to monitor the outcome of the initial engagements and decide whether additional forces must be directed against the target. The ability to direct strategic bombers against appropriate targets is no less a military capability than the ability to aim artillery accurately. The United States did not have the ability so to direct its bombers before World War II, and the development represented a wartime innovation as significant as the introduction of the tank in World War I.

The other Air Forces innovation was the long-range fighter escort. What role in this development was played by the disastrous duels between American bombers and German fighters over Germany in the second half of 1943? Effective long-range escort fighters had not been thought technologically possible before the war, but a new weapon and a new way of using aircraft was developed within the short period of American bombing operations in Europe. The development of the long-range escort fighter by the United States has been advanced as an example of how the shock of heavy casualties can be effective in forcing military organizations to innovate. The explanation of organizational learning set out in earlier chapters emphasized the importance of a definition of a new strategic measure of effectiveness that enables the organization to focus on the relationship between operational performance and the strategic objective in a way that makes possible an evaluation of alternative operational methods. But perhaps such an elaborate learning process is not always necessary. Perhaps a military catastrophe might force organizational responses even without a well-developed intellectual construct that measures relative strategic effectiveness.

THE INVENTION OF STRATEGIC TARGETING: THE PREWAR MUDDLE

The idea of bombing urban and industrial targets as a way of defeating the enemy without first destroying his army received some official notice within the U.S. Army Air Forces at least thirteen years before the beginning of World War II.[1] However, the Air Forces as an institution did not embrace what would become its concept of operations, that is, the targeting of industrial bottlenecks, until after the outbreak of war in Europe in 1939. Components of the Air Forces, most notably the Air Corps Tactical School (ACTS) initially focused on bombing urban populations for the psychological effect it would have on the enemy. Although scholars have found few direct indications that the writings of the Italian theorist of air warfare Emilio Douhet were widely read in the Air Forces, certain ACTS publications reflected Douhet's way of thinking. A 1926 ACTS pamphlet, "The Employment of Combined Air Forces," shared Douhet's emphasis on the destruction of enemy morale and will by means of attacks on his interior, arguing that "terrorizing the whole population of a belligerent country while conserving life and property to the greatest extent . . . is a means of imposing [our] will with the least possible loss by striking vital points rather than by wearing down an enemy to exhaustion."[2]

The concept itself was ambiguous, leaving unclear how enemy populations were to be terrorized by attacks on selected targets that spared life and property. No clear strategic measure of effectiveness proceeded from this concept. Just what would be attacked, and how would the strategic performance of the bombers be measured? Similar ambiguities mark the writings of the leading American champion of air power, Brigadier General William Mitchell. In his *Winged Defense* (1925), he suggests that air power makes the entire civilian population of an enemy vulnerable. This would make the enemy population much less likely to go to war, and much more likely to seek peace if war did break out. But where and how was the enemy

[1] The Army Air Corps was reorganized as the Army Air Forces in 1941. To avoid confusion, I use the term Army Air Forces to refer to both the prewar and the wartime organization.

[2] Thomas H. Greer, *The Development of Air Doctrine in the Army Air Force 1917–1941* (Washington, D.C.: GPO, 1985), pp. 41, 51. See also the references to the April 1928 statements by the Office of the Chief of Staff of the Air Corps that the will of the enemy was the objective that could be attacked directly by the Air Corps without engaging the enemy's armed forces, in Wesley Frank Craven and James Lea Cate, eds., *The Army Air Forces in World War II*, 6 vols. (Chicago: University of Chicago Press, 1948), 1:46.

[150]

vulnerable? What amount of damage, and of what kind, would have to be inflicted before the enemy surrendered? Mitchell wrote nothing about how the U.S. Air Forces should decide what to bomb or what kind of further campaign should be waged if the vaguely defined bombing strategy did not, after all, immediately produce enemy capitulation. His few concrete recommendations had to do with organizing the defense of the United States against enemy attack by sea and air, not with how to bomb a European enemy.[3]

Air Forces doctrine did not become more coherent until the outbreak of World War II. There was a fundamental conflict between the vision of future wars that emerged in one component of the Air Forces, the ACTS, which was guided by the emerging technological ability to construct very long-range bombers, and the clear intent of American foreign policy, which was to eschew capabilities for intervention in European wars. Since such intervention was not legitimate, the strategic mission of long-range bombers had to be confined to assisting in the defense of the United States. Thus, while individuals within the Air Forces, such as the instructors at the ACTS and one commander of GHQ Air Force (the Air Forces command in which bombers were initially placed), General Frank Andrews, did call for the eventual development of intercontinental bombers for use against enemy homelands,[4] the Air Forces as an organization remained hostile to the concept of the offensive use of long-range bombers. Reflecting this, the ACTS in January 1935 formally promulgated a doctrine that relegated bombers to the air defense of the continental United States. They would help prevent the establishment of hostile air bases within striking range of the United States and, failing that, could attack such bases to prevent them from being used. The use of bombers against naval attack and invasion fleets was also emphasized. The B-17, which was first delivered to the Air Forces in 1937, was publicly hailed by GHQ Air Force as "the best bombardment aircraft in existence, particularly for coastal defense." The ability of the B-17 to make a one-way flight to Europe or the Philippines and then to refuel at local bases to attack European or Asian targets could not be denied, but the intent to so use the aircraft was publicly rejected by another commander of the GHQ Air Force, who noted in November 1937 that "our National Policy is defensive, and we do not now consider such possibilities."

[3]William Mitchell, *Winged Defense* (New York: G. P. Putnam's Sons, 1925; rpt., New York: Dover, 1988), pp. 14, 215–19.
[4]Greer, *Development of Air Doctrine*, p. 94.

Army-navy conflicts over which service was to be responsible for the aerial defense of the United States against naval attack were resolved by a law prohibiting the operation of army aircraft based in the continental United States at distances greater than one hundred miles from shore. As Europe moved toward war, the definition of the Air Forces' roles was expanded to include hemispheric defense, but the orientation remained defensive.[5]

The absence of a national political objective consistent with long-range offensive air operations gave a weapon to those officers within the army who wanted to reduce the Air Forces to a combat arm that supported ground force operations.[6] It also provided the rationale for attacks on the bomber program by army officers who, for a variety of reasons, sought to curtail the development of an autonomous long-range bomber force within their service. In 1936, as technical developments brought a true intercontinental bomber closer to reality, the head of army procurement in the G-4 division of the General Staff, George Spalding, submitted a report stating that there was no mission requirement for the B-17 or for any new four-engine long-range bomber. The deputy chief of staff, Stanley Embick, and the Joint Army-Navy Board agreed with Spalding to the extent that they believed that there was no need for a bomber with a range greater than that of the B-17. As a result, all funds for four-engine bomber procurement, including funds for the B-17, were eliminated from the FY 1940 and 1941 military budgets approved by the secretary of war in August 1938. The effect of these decisions was to reduce the number of B-17s actually delivered to the army by September 1939 to fourteen.[7]

---

[5]Craven and Cate, *Army Air Forces in World War II*, 1:49, 1:61–63, 1:68, 1:119.

[6]The Training Regulation 440–15 of 26 January 1926 stated that all air units were "based on the fundamental doctrine that their mission is to aid the ground forces to gain decisive success." The 26 September 1934 Joint Board of the Army and Navy publication "Doctrines for the Employment of GHQ Air Force" stated that force would be deployed to the field with any army expeditionary force and would be used "to operate as an arm of the mobile Army . . . in the conduct of air operations over the land in support of land operations." Cited, ibid., pp. 45, 48.

[7]Greer, *Development of Air Doctrine*, pp. 95–100. These decisions were reversed by President Roosevelt in September 1938, when he called for ten thousand military aircraft, and in September 1940, when he called for fifty thousand military aircraft. Paradoxically, these calls for massive increases in aircraft production did not by themselves favor bomber production. Fighters were smaller and more economical to produce than bombers, so undifferentiated calls for the rapid expansion of aircraft production tended to lead to increases in fighter production and to increases in the production of existing aircraft types, as opposed to those in development. It took vigorous lobbying by Assistant Secretary of War Robert Lovett in 1940 and 1941 to

The procurement decisions were reversed by subsequent presidential action. What was of more long-lasting consequence was the impact of political constraints on the development of strategic targeting. There were advocates of strategic bombing in the prewar Air Forces. Moreover, the idea of selective bombing of industrial targets had long been discussed,[8] and instructors at the ACTS had lectured on this subject.[9] Nonetheless, at the beginning of the war the Air Forces lacked both a developed doctrine of strategic targeting and the necessary intelligence capability. General Hap Arnold, chief of staff of the Army Air Forces from 1938 to 1945, noted in retrospect that the major weakness in his force on the eve of war was "the lack of a proper Air Intelligence organization. . . . Our target intelligence, the ultimate determinate, the compass on which all the priorities of our strategic bombardment campaign against Germany would depend, was set up only after we were actually at war."[10]

Prewar instructors at the ACTS understood the need for an intelligence capability relating to targets. In October 1939, Major Muir Fairchild lectured on the problem of target selection. Which was the most decisive target for bombers: enemy air, ground, or naval forces, a nonmilitary target, or the national infrastructure? Reasonably enough, Fairchild argued that the answer would differ from country to country. The decisive targets in Mexico would be quite different from those in the United Kingdom. Only a study of particular cases would produce useful answers. But the capability for this kind of detailed study did not exist in the Air Forces until 1941. There was not even any routine coordination between the Air Forces and the army G-2 staff for intelligence.

---

reduce FDR's production targets in order to make more heavy bomber production possible. See Jonathan Foster Fanton, "Robert A. Lovett: The War Years" (Ph.D. diss., Yale University, 1978), pp. 73–76.

[8]Ernest Swinton published a short story in 1907 in which aircraft carrying bombs shut down a key railroad bridge, not by striking at the bridge itself but by striking at a pile driver that was vital to its construction. The secondary target was the power plant for the pile driver. The story is reprinted under the pseudonym Ole Luk-Oie in *The Green Curve and Other Stories* (Edinburgh and London: William Blackwood and Sons, 1916). An edition of *The Green Curve* was kept on the nightstand of a British army officer who became an American citizen and a colonel on the staff of the Eighth Air Force planning staff, which in World War II linked the economists doing target planning for the Office of Strategic Services with the Eighth Air Force mission planners. See Walt Whitman Rostow, *Pre-Invasion Bombing Strategy: General Eisenhower's Decision of March 25 1944* (Austin: University of Texas Press, 1981), pp. 17–18.

[9]Heywood Hansell, Jr., *The Air Plan That Defeated Hitler* (Atlanta: Higgens-McArthur, 1972), pp. 24–29.

[10]H. Arnold, *Global Mission* (New York: Hutchinson, 1951), p. 124.

Two factors were responsible for this state of affairs. First, the absence of intelligence about foreign targets reflected the essentially defensive strategic role assigned bombers. While thought was given at the ACTS to target selection, the problems remained hypothetical. Industrial systems were evaluated at the ACTS to determine their vulnerability to bombing and to evaluate the impact of their destruction on the larger economy, but, in the words of one historian, "in most cases the actual study was made of appropriate targets in the United States. . . . Although the instructors were thinking in terms of offensive air actions against potential enemies, they found it more practical and discrete to talk and study in terms of possible air attack on the United States." Second, there was no money for analysis of foreign industries that might be targeted and their relation to the larger economy, so analysts simply assumed that all industrial economies resembled that of the United States. Once this pattern was established, it became easy to identify key industrial targets in the United States and then to make analogies to foreign economies.[11]

In terms of the framework developed in the last chapter, there had not yet emerged in the Air Forces of 1939 a clear measure of effectiveness for long-range bombers. There were competing measures: relating to hemispheric defense, support for ground forces, and a partially defined concept of attacking industrial targets. There was a corresponding lack of intelligence mechanisms that would provide data relevant to the measure of effectiveness to determine whether performance was satisfactory. But this was the peacetime Air Forces. In the period between the outbreak of war in Europe and the beginning of American bombing operations there, did the United States settle upon a measure of effectiveness and develop the intelligence systems necessary to determine wartime performance of long range bombers?

In some measure the answer is no. The American military intelligence system did focus on the air war between Great Britain and Germany in 1940 and 1941, but the operational issue that received the most attention from our military attachés was strategic defense, not offense. Attachés in both London and Berlin reported back on the effectiveness of air-defense systems against bomber attack. The U.S. Army attaché reporting from Berlin in 1940 and 1941, for example, submitted translations of unclassified German literature about

---

[11]Greer, *Development of Air Doctrine*, pp. 53, 81.

the organization of German air defenses and observations concerning German antiaircraft searchlights, ground observers, and pursuit aircraft.[12]

Similarly, the 1941 book *Winged Warfare*, written by Hap Arnold and Ira Eaker, the officer who would command the Eighth Bomber Command of the Eighth Air Force, drew lessons from the first year of air war in Europe, but concentrated on the aspects that pitted one air force against another, not on ground targets. The ability of bombers to penetrate fighter defenses, the need to destroy enemy air forces as the necessary first phase in any aerial campaign, and the technological balance between the American, British, and German air forces came in for much discussion. The question of how to select targets the destruction of which would cripple the German war economy came in only for the kind of general discussion it had received at the ACTS before the outbreak of the war. Factories of all kinds, electrical power grids, critical transport links, and even centers of finance were all suggested as plausible targets. But no priorities were suggested, nor any method for gathering and using wartime intelligence to determine priorities. The authors of *Winged Warfare* simply asserted that "every factory in the enemy territory which is producing a vital war material or an essential item of equipment can expect air raids."[13] To be sure, one should not expect detailed intelligence analysis in a popular book. Yet the general doctrine of the U.S. Army Air Forces for air-to-air combat was laid out in the book, while the concepts that would guide strategic bombing were left unclear.

THE INVENTION OF STRATEGIC TARGETING: WARTIME INNOVATION

If the U.S. Army Air Forces had been slow and confused before the war, it displayed admirable speed and energy in wartime and quickly developed the capacity to select and evaluate strategic targets. Beginning in August 1941 and continuing through the summer

[12]See Army attaché reports #17,046, January 8, 1940; #17,465, August 15, 1940; #17,586, September 26, 1940; #17,857, January 3, 1941, all in Record Group 165, Records of the War Department General and Special Staffs, Regional File 1922–1944, National Archives, Suitland Station, Md.

[13]H. Arnold and Ira Eaker, *Winged Warfare* (New York: Harper and Row, 1941), pp. 8, 130, 132.

of 1944, at least four separate organizational efforts were mounted to develop the strategic targeting function. All involved, to differing degrees, attempts to define strategic measures of effectiveness that related bomber operations to the overall course of the war, as well as efforts to provide the data relevant to the measures of effectiveness, that is, to determine how much strategic damage bombing operations were actually causing. The products of each of these efforts came to bear some resemblance to each other, and each came to grief in the same manner. Well-defined measures of effectiveness and routines for utilizing intelligence were quickly developed but functioned imperfectly in practice. Although Ultra intelligence provided unusually good information about the enemy military, it did not provide reliable data relevant to strategic targeting.

The four targeting functions created were in the Air War Plans Division (AWPD) of the Air Corps Staff, the Economics Objectives Unit (EOU) of the Office of Strategic Services, operating out of the American Embassy in London, the Army Air Forces' ad hoc civilian-military group referred to as the Committee of Operations Analysts (COA), and the elements of the British Ministry of Economic Warfare (MEW) that provided formal and informal support to the U.S. Air Forces in the European theater of operations. As in the case of the tank, initial recognition of a need and military action in response were prompt. The Air War Plans Division predated the war, as did the Ministry of Economic Warfare. The origin of the EOU was described in that unit's war diary: "[It] came into being because of a clear gap in the organization of the American Air Staff in Europe. If precision bombing was to be carried out against Germany it was apparent from as early as the summer of 1942 that intelligence would have to be analyzed and organized in forms not already provided by the various British intelligence services. The staff did not exist within the air force to carry out the kind of technical studies envisaged, and civilian aid was invoked." Two Air Forces colonels concerned with target planning took the initiative and approached the OSS for support in April 1942, and a formal Eighth Air Force request for support gave the EOU formal bureaucratic responsibility and standing in September 1942.[14] The COA was created by a De-

---

[14]*War Diary,* Research and Analysis Branch, Office of Strategic Services, London (hereinafter *OSS War Diary*), vol. 5, *Economic Outpost with Economic Warfare Division,* National Archives, Washington, D.C., pp. 1, 11, 16. This diary was written shortly after the end of the European war by a member of the staff of the EOU, Walt Rostow, and Rostow drew heavily on it for his 1981 book *Pre-Invasion Bombing Strategy.* Where

cember 1942 request by Hap Arnold; the initiative was motivated by a need for intelligence, in time for the January 1943 Casablanca Conference, concerning how the bombing campaign could be conducted so as to weaken the German military enough to permit the invasion of France in May 1944.[15]

The Air War Plans division produced the first set of target priorities in August 1941 in Air War Plan 1 (AWPD-1). In the words of one of the planners involved, as of yet "there were no commonly acceptable formulae for such things as: (1) the methods to be employed in the air offensive, (2) the specific objectives to be sought, (3) the targets to be attacked." Less still was there a common understanding of the strategic relationship of the air war to the rest of the war. Notwithstanding the lack of any method to evaluate the relative value of alternative target plans, AWPD-1 recommended that the German electrical power, transportation, and oil systems be attacked. In the absence of a strategic measure of effectiveness, the makers of AWPD-1 did not recommend on the basis of an analytical effort that compared the impact of bombing military forces, various industrial complexes, transportation systems, or producers of primary products. The emphases of the first target plan appear in large measure to have resulted from the accidental and selective availability of intelligence about Germany. One of the first recruits to the AWPD staff was a civilian businessman, Malcolm Moss, who pointed out to the army officers that American banks had financed the construction of the German electrical power system, and were likely to have in their files the blueprints of power plants and electrical distribution grids that had accompanied the German loan applications. This proved correct. Similarly, another member of the staff had been an oil man who had worked at the Romanian oil refineries at Ploesti, and he argued that these and the German synthetic oil plants were both critical and vulnerable. Less was known about the German rail network or light metals industry, but enough was known to include them on the target list.[16]

Prewar civilian data about selected industries, rather than a systematic understanding of German industries, their relation to the overall German economy, and the relation of the economy to the

relevant, the diary is referred to on the assumption that it more accurately reflects the thinking of Rostow before he could have been affected by postwar events such as the publication of the *U.S. Strategic Bombing Survey* and the memoirs of Nazi officials.

[15]Craven and Cate, *Army Air Forces in World War II*, 2:353; Perera, *History of COA*, p. 15.

[16]Hansell, *Air Plan*, pp. 51–52, 81–85.

German capacity to make war, appear to have had considerable impact on the target planning. Such civilian data could be extremely useful. Lloyds of London, for example, had written the insurance on the Renault automobile factory at Billancourt that had been captured by the Germans and had detailed plans that greatly aided assessment of the March 1942 bombing raid on that factory by the Eighth Air Force.[17]

But these data had significant limitations when applied to the task of analyzing German strategic vulnerabilities. To evaluate alternative targets, analysts needed more than data about isolated factories or electrical power systems. If the strategic objective was reducing the German capacity to make war, it was not enough to know that Romanian oil refineries were vulnerable to high explosives dropped by bombers. How soon and to what extent the reduction in refinery output would affect German military operations depended on the size of German oil reserves, current rates of military consumption, and the extent to which civilian consumption could be curtailed. Similarly, the impact of bombing the railroad system of Axis Europe would depend on how much excess capacity was built into that system, that is, how much greater the capacity of a system constructed for civilian needs and civilian work schedules would be compared to the minimal wartime needs of a German military utilizing the railroad around the clock and exploiting captive labor working wartime shifts when necessary to make repairs. Such data was not available to the AWPD in 1941.

The Enemy Branch of the British MEW had focused on the German economy during the 1930s and had intensified its intelligence activities in this area with the outbreak of the war. By 1941, however, it had not managed to improve its abilities to collect or analyze raw data. Although the MEW, unlike the AWPD, had had mechanisms for collecting economic intelligence in peacetime, these did not prove appropriate to wartime needs. Before the outbreak of the war in 1939, the MEW had relied primarily on unclassified German documents and contacts with businessmen. When these sources dried up in 1939, new sources did not spring into being. In the words of the official history of British wartime intelligence, "Given the shortage of first hand intelligence, there was a natural tendency to carry over into war-time conditions assessments based on pre-war statistical data and to rely on 'common sense' conclusions." Like the

[17]Ursula Powys-Lybbe, *The Eye of Intelligence* (London: William Kimber, 1983), p. 155.

AWPD, the MEW suffered not only from a lack of data, but from a lack of a clear definition of the problem they were trying to solve, of just what it was about the German economy they needed to know in order to evaluate the strategic impact of bombing. The official history notes that "Enemy Branch did not merely lack reliable intelligence about economic policy and economic administration in Germany; it did not recognize that the study of those subjects was central to intelligence work on economic problems."[18] In terms of the criteria I have set forth previously, the initial British and American efforts to perform a new function, strategic target planning, were not successful because they did not define the way in which strategic bombing was to affect the outcome of the war. While both assumed generally that bombing industrial targets would hurt the Germans, neither had a strategic measure of effectiveness that defined how military operations were related to strategic objectives and indicated how well those operations were progressing toward the strategic objective. Beyond that, both the British and American efforts lacked reliable wartime data on which to base assessments.

These weaknesses became more acute when German oil refineries and oil reserves became a target. What was the relationship between the operational objective of successfully striking at those targets and the strategic objective of hindering the German war machine? Was the oil target a more or less lucrative one relative to the alternatives? The capabilities of the strategic targeting function would be tested by this question. British intelligence and policy committees did recommend that German oil targets be bombed, on the general assumption that oil was in tight supply in the Nazi empire. Oil targets in the Ruhr were struck by the British Bomber Command from October through December 1940. But after the oil targets had been struck, the natural question was posed. How effective had the strikes been at reducing the output of the refineries and synthetic oil plants, and strategically, at reducing the ability of the German military to operate? The answer to the strategic question was unknowable since the British intelligence committees admitted that they knew little about enemy oil reserves, rates of consumption, or the output of synthetic oil plants. An effort was made to answer the operational question, and the conclusion was reached that German synthetic oil production had been reduced by 15 percent. What is

[18]F. H. Hinsley, *British Intelligence in the Second World War: Its Influence on Strategy and Operations*, 3 vols. (London: HMSO, 1979), 1:64–65, 1:227, 1:246.

striking about this estimate is the process by which it was reached. The intelligence committee broke the synthetic oil production system into subprocesses, assessed the relative importance of each subprocess, estimated the number of bombs needed to completely halt each, along with the amount of time needed to repair it, and calculated the probability of damage done to those subprocesses by notional, not actual, bombing raids of various sizes. A model of the oil system was built and imaginary attacks were run against the model. The judgment of the official historians of this intelligence effort is harsh: "Acknowledged by its author as an experimental and theoretical approach to the problem, and depending on data which owed virtually nothing to intelligence, the method thus yielded the highly optimistic estimate of bomb damage."[19] The British target function that was supposed to provide indications of how well strategic bombing was performing could not do so reliably at either the operational or strategic level. In comparison with this effort, the AWPD failings seem less egregious. The American strategic analytical effort did not noticeably improve during the first year in which American bombers were employed in attacks against Europe. In large part this was because American bombers initially raided German U-boat bases in France, simply because the U-boat threat was so severe that any and all weapons were turned against it.[20] By the end of 1942, however, senior military leaders returned to the question of whether and how American bombing might have a strategic impact on the war. The Committee of Operations Analysts took a fresh look at German economic targets, and the oil industry and ball-bearing factories in particular, and submitted an interim report in January 1943, before the Casablanca Conference. The COA had recruited an impressive list of businessmen and academics, including the Princeton economist Edward Meade Earle, the Harvard economist Edward Mason, and Wall Street bankers and lawyers Elihu Root, Jr., Thomas Lamont, and Fowler Hamilton.[21] It also drew on the knowledge of America industrialists, who were asked to evaluate the ability of their own industries to absorb damage from enemy bombers.[22]

These men, along with a staff drawn from the officers serving in Air Forces' Directorate of Management Control, recommended that

[19]Ibid., 1:241, 1:243, 2:131, and appendix 12, "The Oil Target," pp. 2:691–92.
[20]Craven and Cate, *Army Air Forces*, 2:313.
[21]David MacIssac, *Strategic Bombing in World War II: The Story of the United States Strategic Bombing Survey* (New York: Garland, 1976), p. 25.
[22]Craven and Cate, *Army Air Forces*, 2:352.

the Ploesti oil refineries and the synthetic oil plants be bombed. The COA however, could not present a convincing case that such an attack would have an immediate impact on German military operations because it still did not have reliable information about the size of German reserves. The COA's report stated: "Existing stocks of all petroleum products are estimated variously from 2.4 to 6.0 million tons. Wide variation of figures [is] inevitable due to impossibility of direct observation." Its recommendation that oil be made a target was the product less of analysis than of broad observations and assumptions—that Germany had not anticipated the war with the Soviet Union would take so long or use so much oil, that German air activity against Great Britain and over North Africa seemed to have been curtailed, and that the Germans seemed to be preoccupied with Romania and the oil producing areas of the Caucasus in the Soviet Union.[23]

British and American analysts disagreed at the beginning of 1943 about the size of the German oil reserves because they could not agree on the size of the reserves that Germany took with her into the war. The American analysts estimated that the current German oil stockpile was five million tons; the British, three million. The British further estimated that the German oil system was stretched tight because it took three million tons of oil to keep the oil pipelines and distribution network full. Any reductions in oil production would force the German to start emptying the pipelines. The American estimate implied that the Germans had a sizeable cushion and that oil production could drop without any immediate impact on the German system. This was an important strategic disagreement that affected judgments about the value of raids against the oil target. Instead of being resolved analytically, however, the disagreement was resolved bureaucratically, by splitting the difference and simply asserting that for planning purposes German reserves would be set at four million tons.[24] The size of the German oil "cushion" was by arbitrary decision made larger than the needs of the distribution system. The 21 January policy directive proceeding from the Casablanca Conference appears to reflect the analysis that the German oil industry had some slack and that bombing it would not produce immediate strategic results. In that directive, the German oil industry was placed fourth on the list of target priorities, behind subma-

[23]Perera, *History of COA*, p. 15, and tab. 7, "Analysis of Western Oil Axis Industry as Bombardment Target."
[24]Hinsley, *British Intelligence*, 2:134, 2:136–37.

rine construction yards, aircraft industries, and transportation.[25] Lack of adequate peacetime and wartime data affected the strategic target function in ways that had significant impact on major strategic choices.

Poor wartime data also played a pivotal role in choosing and then assessing the damage done to one of the most famous targets of American bombers in World War II, the ball-bearing factories at Schweinfurt. The history of the COA suggests that it was an accidental social contact between one of the analysts on the staff of the COA and an American ball-bearing manufacturer that resulted in the complex's being placed on the target list. The industrialist pointed out the extent to which Schweinfurt dominated the German ball-bearing industry.[26] The War Diary of the EOU, however, states that it was a 1943 analysis of the serial numbers of the ball bearings in the engines of German aircraft shot down during the Battle of Britain that showed the extent to which one factory dominated German production.[27]

Where the COA did better than the AWPD was in the articulation of a true measure of effectiveness for the strategic bombing of industrial targets. The committee laid out six criteria for evaluating an industrial system as a target. These compared the total production capacity of a given industry to the minimum enemy requirements for that commodity to determine surplus industrial capacity. The industry was then examined to determine which plants were the most critical in terms of production and most difficult to replace if damaged. The last criterion was the amount of time that would have to elapse before the destruction of a plant by bombing would impair the effectiveness of enemy forces in the field.[28]

The most comprehensive measure of effectiveness for strategic bombing and the most sophisticated innovation in the field of targeting, however, was developed by the EOU during the winter of 1942–1943. The EOU began by preparing Aim Point reports, which designated the critical plants, buildings, and pieces of machinery within a given German industry. This was a more limited, technical task than the selection of strategic targets, since the reports did not

---

[25]Craven and Cate, *Army Air Forces*, 2:305, 2:360. It should be noted, however, that submarine construction yards were placed first on the target list over the objections of the COA and the Eighth Air Force, which doubted the ability of bombers to strike effectively at the German submarine yards. The persistent threat of U-boats to Atlantic convoys made it politically impossible to downgrade this target.

[26]Perera, *History of COA*, pp. 18–19.

[27]*OSS War Diary*, p. 125.

[28]Craven and Cate, *Army Air Forces*, 2:350.

ask whether the industry itself was the system most worth attacking. It was closer to an industrial analysis of plant vulnerabilities. Even so, the task of assembling enough empirical data to yield useful recommendations was extremely difficult under wartime conditions. Although all sources of intelligence were exploited, it is clear that the vulnerabilities of German factories were determined by examining British factories in the same industry and then assuming that German factories had not only the same physical characteristics, but also the same damage control procedures.[29]

The EOU quickly moved beyond this technical aspect of targeting to a more ambitious task, the development of a theory of choice for selecting strategic targets. In memoranda circulated from December 1942 until March 1943, a set of criteria was elaborated. Many of the criteria were self-evident. A good target had to be important for war production, its destruction would have to lead to general effects on the war economy, it had to be in a sector in which there was little or no surplus capacity or for which there were no easily developed substitutes, and it had to be physically vulnerable to bombing. Impact on the enemy had to be compared to the number of bombers that would be lost in the attack. The criteria were refined by the addition of a metaphor. Enemy forces in the field, it was argued, could be regarded as a pool into which production flowed and from which combat losses were taken. A category of weapons for which the pool was very large relative to the flow of production was a poor target, since even the complete halt of production would leave a large pool of weapons in the field that would have to be destroyed by direct combat. It would be preferable in these cases to devote effort to increasing combat losses by the enemy. German submarines had this character since there were many more U-boats at sea, in port, or in training than were delivered from submarine yards each month. Industries in which production was large relative to the pool, however, made a more lucrative target for bombing, since even large German losses in the field would be quickly made good by production, while destroying the source of the weapons would have lasting impact. German fighter aircraft, it was argued, fell into this category.[30]

Crucial to this measure of effectiveness was an element of timing. Any bombing of an industrial target would have some effect on German capacity to wage war. What was important was when those ef-

[29]*OSS War Diary*, pp. 20–21, 25, 63.
[30]Ibid., pp. 32–36, 39.

fects would be felt by the German forces in the field. This was the critical question, because the central mission of the strategic bombing campaign, under the terms of the POINTBLANK directive issued by the Casablanca Conference, was the weakening of the German military in time to affect the outcome of the planned invasion of France in the spring of 1944. The EOU had been specifically asked by the Air Forces about the relationship of the bombing campaign to the optimal moment for invasion.[31] Bombing that would have an effect after the invasion was less critical.

For precisely this reason, the German oil industry was rejected as a target by Supreme Allied Commander Dwight Eisenhower. In the critical meeting of 25 March 1944 to decide preinvasion bombing strategy, the favored target of the EOU was rejected because EOU's own analysis showed that attacks on enemy oil capacity would show effects only three months after the bombing campaign, too late to affect the critical early weeks of the invasion.[32] In this case, however, it was more important to have reliable wartime data than a sophisticated measure of effectiveness. When Eisenhower relented and allowed the Air Forces to bomb German oil targets in May, the effects were immediate, and not delayed by three months as suggested by EOU analysis. The systematic bombing of oil targets began 12 May 1944.[33] On 13 May Ultra intercepts of Luftwaffe communications revealed that German antiaircraft units were urgently being relocated to protect synthetic oil plants, even at the price of depriving German fighter-aircraft factories of their protection. Luftwaffe communications on 14 May reported that motor transport units, including those in combat units, were ordered to switch their fuel from petroleum to wood gas. And on 5 June, the German Air Forces Operation Staff in Berlin broadcast this message: "As a result of renewed encroachment into the production of aircraft fuel by enemy action, the most essential requirements for training and carrying out production plans can scarcely be covered." A German major taken prisoner in France confided to another prisoner in a monitored conversation: "Above all, we have no petrol left. We can no longer move any numbers of troops by means requiring petrol; only by rail or by marching on foot." Throughout June and July, Luftwaffe com-

[31]Ibid., p. 41.
[32]The minutes of the meeting at which this decision was made are reprinted in Rostow, *Pre-Invasion Bombing Strategy,* appendix A, pp. 88–98.
[33]Craven and Cate, *Army Air Forces,* 3:174–76.

munications would reveal the curtailing of aerial operations as a result of fuel shortages.[34]

The EOU analysis had been badly mistaken. The strategic effectiveness of the U.S. heavy bombers may have been hindered rather than helped by the wartime innovation of the strategic targeting function. What had happened? The absence of adequate wartime data that would have enabled analysts to understand the actual, as opposed to theoretical, vulnerabilities of the German war economy appears to have been important. Despite the development of a sophisticated measure of strategic effectiveness for bombing, the OSS War Diary notes that EOU analysts relied on the reports by the British and American intelligence committees for their figures on the German oil industry and oil reserves that were used to compare alternative target sets.[35] Those committees had manufactured the estimates about German oil reserves from sketchy knowledge, theory, and bureaucratic compromise. Bad data concerning reserves resulted in the erroneous conclusion that the Germans had an oil "cushion."

Similarly, lack of knowledge of the realities of the German oil industry had forced EOU analysts to assume that British oil refineries were analogous to their German counterparts. As postwar reports showed, however, the German synthetic oil plants constructed in wartime were much more integrated into the synthetic rubber and chemical industries than were the comparable British plants, and attacks on the German plants destroyed petroleum, rubber, and chemical manufacturing capacities simultaneously.

A shortage of reliable wartime data also hindered the strategic targeting function in assessing the impact of Allied bombing on the German ball-bearing industry in 1943. Placing Schweinfurt on the target list was only half of the process. Determining the impact of the bombing whether it justified the loss in bombers was the other half. The strategic objective behind the bombing of Schweinfurt was the crippling of the German aircraft engine industry. To what extent did the raids in August and October 1943 actually accomplish this

[34]National Security Agency, SRH-013, "Ultra History of U.S. Strategic Air Force Europe vs. German Air Force," June 1945, Record Group 457, National Archives, Washington, D.C., pp. 178–80, 190, 199, 211, 212–13. This report was reprinted in 1980 under the same title by the University Press of America. See also SRH-015, "Notes on German Fuel Position: G-2 SHAEF Memoranda and Studies Concerning the Use of Ultra in Evaluating German Fuel Position," 31 March 1945, Record Group 457, National Archives, Washington, D.C.

[35]*OSS War Diary*, p. 71.

objective? An October 1943 report by the British MEW sent to the U.S. War Department General Staff attempted such an assessment. It is notable for the uncertainty of its conclusions. It noted that numerous hits were photographed for all the factory buildings, although the main productive machinery appeared not to have been hit. The actual loss of production, the report noted, would depend on German reconstruction, but production might be cut by 16.5 to 25 percent for a period of six months. A fundamental error was to assume that there were no significant stockpiles of ball bearings. Even so, the report warned that the actual impact of the bombing on military production would "depend upon whether a bottleneck has been created in any particular line of production, of which there is no present evidence."[36]

Did German aircraft production suffer after the ball-bearing factories were bombed? The Research and Analysis Branch of the OSS tried to answer this question in March 1944. It noted a steady decline in German single-engine fighter production beginning in the first quarter of 1943 and continuing through the bombings of Schweinfurt.[37] This finding proved to be incorrect. After the war, it was discovered from German records that while single-engine fighter production did temporarily decline when bombing forced the ball-bearing factories at Schweinfurt to relocate, the general trend in fighter production in both halves of 1943 was upward, not downward.[38] In fact, postwar analysis showed that "no type of German military production was appreciably curtailed for a lack of bearings."[39] What had produced this gross failure in the strategic targeting mechanisms?

As noted earlier, when the war started, Allied intelligence was deprived of much of the data on which it had based its peacetime estimates of German weapons production. Despite the rapid growth

---

[36]"Industrial Damage Report No. 76" covering the month of October 1943, 8 January 1944, Objectives Department, Ministry of Economic Warfare to Military Intelligence Division, War Department General Staff, transmitted via Major H. M. Stout, Assistant Military Attaché, London, 25 January 1944, pp. 19–20, National Archives, Suitland Station, Md.

[37]"Preliminary Assessment of Effects of USSAFE Attacks of 20–25 February 1944 on German Aircraft and Anti-Friction Bearing Industries," Research and Analysis report #1973, OSS, 11 March 1944, Record Group 165, Regional File 1922–1944, National Archives, Suitland Station, Md., graph following page 2.

[38]Rostow, *Pre-Invasion Bombing Strategy*, pp. 26–27, citing United States Strategic Bombing Survey findings; Williamson Murray, *Strategy for Defeat: The Luftwaffe, 1933–1945* (Maxwell AFB, Ala.: Air University Press, 1983), pp. 105, 230.

[39]Burton Klein, *Germany's Economic Preparations for War* (Cambridge: Harvard University Press, 1959), p. 232.

in the intelligence establishment during the war the quality of the data about the German economy remained poor. For example, at the end of 1943, one-third of all data about Germany was still being gleaned from unclassified publications or from business travelers in touch with Germans outside Germany.[40] The impressive American and British communications interception and cryptanalysis effort known as Ultra was the main source of communications intelligence utilized in the analysis of the strategic bombing of Germany, since the U.S. Air Forces did not try to replicate the British effort in Europe.[41] One American participant in the Ultra operation, Supreme Court justice Lewis F. Powell, has stated without elaboration that Ultra was of importance in shaping the operations and plans of the U.S. Strategic Air Forces in Europe.[42] A close look at recently declassified documents suggests, however, that Ultra revealed surprisingly little about the German economy and so provided very little help in assessing the impact of strategic bombing.

A study of the impact of Ultra on strategic target planning made immediately after the war noted that Ultra intercepted operational military traffic, not the reports of factories. In the words of the report, "Even the production of such items as aircraft and tanks was not normally within the province of MSS Ultra." Reports on the output of ball bearings or aircraft engines would seldom find their way into military radio traffic. Factory managers communicated by telephone, which could not be intercepted in the same way. As the war progressed, however, the destruction of telephone lines forced the German police to rely on military radios, and police reports would often include some references to the effects of a bombing raid. Later in the war, German industry was put under a military command that coordinated weapons production, and Ultra intercepts of its radio traffic did provide some data on industrial output. But both these sources yielded data "so fragmentary that it was of little use."

Typical was a report on the effect of the bombing of Schweinfurt that was broadcast over German air force radios and was intercepted and decoded. It reported "considerable damage to ball bearing factories. One shadow plant completely destroyed." How was this to be assessed by the Allies? What was "considerable damage?" The destruction of a shadow plant, a secondary plant that supplemented

[40]Hinsley, *British Intelligence*, 3:53–54.
[41]Ibid., 2:49.
[42]Diane Putney, ed., *ULTRA and the Army Air Forces in World War II: An Interview with Associate Justice of the U.S. Supreme Court Lewis F. Powell, Jr.*, Office of Air Force History, United States Air Force, Washington, D.C., 1987, pp. 35–38.

the production of the main factory, was important. But was that particular shadow plant one that had been known to the Allies or was it a source of production that had not been incorporated into Allied production estimates? Because of this and other problems, the postwar report concluded that "generally speaking . . . Ultra was not a source of major importance in the initiation of target systems for attack." The report further concludes: "On the whole, it is fair to say that the major decisions on the employment of strategic air power would have been the same had Ultra not been available." Ultra did provide valuable confirmation of the value of bombing German oil, but had not contributed to the decision to initiate that bombing, nor had it provided similar after the fact confirmations in other industrial areas.[43]

In the case of ball bearings and German aircraft production, it was not Ultra that provided the Allied data on production. The only reliable method for assessing German aircraft production was to analyze the serial numbers of aircraft and aircraft components, including ball bearing and engines. Such serial numbers were collected from all types of captured and destroyed enemy weapons by the joint Anglo-American General Intelligence Unit. The technique yielded the first reliable estimates of German weapons production, which were used, for example, to reduce estimates of German prewar tank production from forty thousand to fourteen thousand. There were problems with this kind of analysis. Only captured or destroyed enemy weapons could be studied, so certain weapons types escaped analysis. This included German fighters, such as the FW-190, that only flew over occupied Europe, precisely the aircraft the U.S. Air Forces were most interested in.[44]

Another problem was deliberately introduced by the Germans. Beginning in the spring of 1943, and unknown to the Allies until after the war, the Germans introduced random gaps into their sequences of serial numbers. From that point on, Allied production estimates ceased to be reliable.[45] The spring of 1943, by unhappy coincidence, also marked the time when the U.S. Army Air Forces began to switch away from attacks on U-boat targets and toward the German war economy. By the time the German aircraft industry became a target, reliable data for assessing the impact of bombing was no

---

[43]"Allied Strategic Air Force Target Planning," August 1945, declassified by the National Security Agency, SRH-017, Record Group 457, National Archives, Washington, D.C., pp. 2–3, 12–13.

[44]*OSS War Diary*, pp. 63, 118, 123.

[45]Hinsley, *British Intelligence*, 3:63.

longer available. As a result, the U.S. Army Air Forces incorrectly concluded that they were winning the war against the German air force by attacking its production base.

The final assessment of the success of the effort to direct U.S. strategic bombing must be sobering. Wartime innovation by the OSS did produce a clear analytical mechanism for selecting targets for the strategic bombers. Organizational routines for translating operational reports and data in terms of that strategic measure were in place by 1943. But the absence of good data, either because of mirror imaging (projecting one's own ideas onto the enemy), deceptive enemy practices, or the inappropriateness of collection techniques, meant that the selection of targets tended not to be guided by realities about the German economy. In the judgment of the official history, the new strategic targeting mechanisms proceeded in the dark, guided by preconceptions and intuition: "There existed in almost every instance a shortage of reliable information, and the resulting lacunae had to be bridged by intelligent guesswork and the clever use of analogies. In dealing with this mass of inexactitudes and approximations the social scientist finds himself in no special position of advantage over the military strategist or any intelligent layman, and an elaborate methodology may even, by virtue of a considerable but unavoidably misguided momentum, lead the investigator astray."[46]

In terms of our analysis of wartime innovation, the development of strategic targeting represents a complicated case. The innovation was successful in that it was quickly adopted and integrated into operational and intelligence systems. It successfully defined what it was that needed to be learned in wartime, thereby solving, in a conceptual sense, the problem of learning how to learn in wartime.

But it was not a success in terms of performance goals, which were to optimize the use of strategic bombing. One analysis of Allied bombing of Germany based on a review of the data collected by the postwar U.S. Strategic Bombing Survey persuasively argues that bombing aided the allies most by forcing the Germans to devote one-third of their war production to antiaircraft munitions of all types—fighter aircraft, radar, antiaircraft artillery and ammunition. To give some sense of the scale of this diversion of resources, the value of German production for antiaircraft purposes in June 1944 was greater than the entire German war production on the eve of

---

[46]Craven and Cate, *Army Air Forces*, 2:369.

the invasion of the Soviet Union.[47] There is no sign that the British or American target selection and analysis functions were sensitive to this aspect of the bombing either before or during the major bombing campaigns. This must be counted as a failure in the implementation of the innovation.

This case provides a useful warning. Even with modern intelligence collection and analysis techniques that performed relatively well, the business of getting good information in wartime remained full of uncertainties and unknowns. The errors of the men and women doing the work of target selection in world War II cannot be attributed to any gross bureaucratic or intellectual failures. They resulted from tremendous difficulties of getting information in the midst of war.

INVENTING THE LONG-RANGE ESCORT FIGHTER

If the development of the targeting function serves to emphasize the difficulty of obtaining relevant intelligence in wartime, the innovation of the long-range fighter highlights the way in which wartime innovation can be blocked or facilitated by enemy countermeasures. The crucial innovation for the development of the long-range escort fighter, the introduction of the drop fuel tank, was initially blocked by the reasonable expectation that simple countermeasures could render the drop tanks useless. Bomber missions were conducted without full escorts, and this had the unanticipated effect of leading the enemy to give up the countermeasures against drop tanks. For this reason, this innovation is not an example of organizational learning so much as a case in which initial wartime actions shaped the behavior of the enemy so that innovations believed to be impossible in peacetime became feasible. It was not the shock of combat losses that prompted this innovation, but the dynamic combat situation that made innovation possible.

Prewar thinking within the U.S. Army Air Forces about the need for escort fighters suffered from the ambiguity and uncertainty that surrounded the entire subject of long-range strategic bombing. General Elwood "Pete" Quesada, a fighter pilot who ultimately became the commander of the Ninth Fighter Command that operated over France during the Normandy landings, reflected on the absence of settled doctrine before the war concerning how U.S. bombers would

[47]Klein, *Germany's Economic Preparations*, p. 233.

penetrate enemy air defenses: "There was almost an ignorant disregard of the requirement of air superiority. It was generally felt, without a hell of a lot of thought being given to it, that if there should occur an air combat . . . it would occur at the target."[48] The ACTS argued during the 1920s and 1930s that long-range daylight bomber missions would have to have fighter escorts, although fast, heavily armed high-flying aircraft might be able to penetrate enemy air defenses unescorted. If escorts were not available, the bombers would have to fly at night.[49]

The first months of the European war led to an Air Corps Board study that concluded that fighter escorts for long-range bombing missions inside enemy territory were necessary: "If unsupported bombers encounter a reasonable force of heavy fighters they will, in all probability, suffer severe losses." The study then muddled its conclusion by stating that the absence of fighter escorts would not justify a decision not to bomb an important target.[50] The chief of staff of the Air Forces, Hap Arnold, wrote on the eve of American entry into World War II that the experiences of the Battle of Britain "showed conclusively that the only reliable antidote to the enemy bomber is the fighter" and the converse, that fighter escorts would be needed "to accompany bombers on long missions into enemy territory."[51]

Air Forces doctrine before the first U.S. missions over Europe was generally supportive of escort fighters. Nonetheless, the long-range escort capabilities of the Air Forces were limited. This contradictory state of affairs can be explained by the state of aviation technology at the time. Given the technology of the 1930s, a fighter with a range long enough to make it useful for long-range escort service would, it was believed, have to be a large, multiengine airplane. It would be maneuverable enough to handle enemy bombers, but not so agile as the short-range fighters that U.S. bombers would encounter. A long range-fighter could be designed, but it would be good only for air defense of the continental United States. This was, in fact, the mission for which the P-38 Lightning was designed, and in 1941 Ameri-

[48]Interview in Richard H. Kohn and Joseph P. Hanrahan, *Air Superiority in World War II and Korea: An Interview with Generals James Fergusson, Robert M. Lee, William Momyer, and Lt. General Elwood R. Quesada* (Washington, D.C.: GPO, 1983), pp. 14, 18.
[49]Kenneth P. Werrell, "The Tactical Development of the Eighth Air Force in World War II" (Ph.D. diss., Duke University, 1969), p. 7; Bernard Lawrence Boylan, "The Development of the American Long Range Escort Fighter" (Ph.D. diss., University of Missouri, 1955), pp. 21–22.
[50]Boylan, "Long Range Escort Fighter," pp. 41–42.
[51]Arnold and Eaker, *Winged Warfare*, pp. 7–8.

can pilots doubted its ability to handle enemy fighters.[52] Another large multiengine airplane for the fighter escort mission was developed, the YB-40, a variant of the B-17 that carried extra armor and guns. This airplane, however, had to be abandoned in May 1943 when, despite persistent efforts, it proved to be too slow and its guns little more effective than the standard armament of the B-17.[53] In pursing this technical line of development, the U.S. Air Forces were utilizing the best available technology and the explicit advice of the British, which was based on their wartime experiences. In response to the Royal Air Force request for a long-range fighter, Air Vice Marshall W. S. Douglas wrote in March 1940: "It must, generally speaking, be regarded as axiomatic that the long-range fighter must be inferior in performance to the short-range fighter. . . . The question had been considered many times, and the discussion has always tended to go in circles. . . . The conclusion had been reached that the escort fighter was really a myth. A fighter performing escort missions would, in reality, have to be a high performance and heavily armed bomber."[54]

Thus while it was believed that an effective long-range escort fighter would be extremely useful if it were practical, there was little prospect that one would be available for use soon; in the meanwhile, the self-defense capabilities of heavy bombers would have to be maximized through the use of large formations that could mass the firepower of the individual bombers. Arnold wrote the chief of the Air Staff of the Royal Air Force in April 1942, four months before actual U.S. bombing began, that "it is possible that with the greater defensive power of our bombers, and a carefully developed technique of formation flying with mutually supporting fire, that our bombers may be able to penetrate in daylight beyond the radius of the fighters."[55]

---

[52]Boylan, "Long Range Escort Fighter," pp. 33, 62, 77. Postwar judgments about the effectiveness of the P-38 as an escort fighter vary. When used over North Africa and Europe in this role, it clearly helped to reduce bomber losses, but Adolf Galland, the German commander of the fighter units defending Germany against American bombers, denigrated the capabilities of the P-38, citing the same shortcomings American fighter pilots had pointed to in 1941—its large size and relative lack of maneuverability. Galland attributed the apparent success of the P-38 as an escort fighter to German operational errors and the quantitative superiority of the American Air Forces. See Adolf Galland, *The First and the Last: The German Air Force in World War II* (Champlin, Ariz.: Champlin Museum Press, 1986), pp. 208, 210.

[53]Craven and Cate, *Army Air Forces*, 2:655, 2:680.

[54]Murray, *Strategy for Defeat*, p. 131; Boylan, "Long Range Escort Fighter," pp. 47–50.

[55]Boylan, "Long Range Escort Fighter," p. 75.

This decision did not mean rejection of fighter escorts. The Army Air Forces were ambivalent. Escorts were desirable, but senior leaders had confidence that strategic bombing without escorts was possible. This ambivalence was reflected in both plans and operations. The major program for the strategic air campaign over Europe developed by the U.S. Army Air Forces in August 1942, AWPD-2, provided for an increase in the number of fighter groups to be stationed in Europe but no provision for strategic escorts and a claim that bombing operations without escorts were "perfectly possible . . . without excessive loses."[56] In practice, escorts would be provided to the limit of the ranges of fighters based in Great Britain. The RAF was asked to provide escorts for American bombers flying shorter range missions into occupied France. When more P-38s were available, and with time for better training for American fighter pilots, they would take over this escort mission from the British. American commanders were understandably reluctant to throw green American fighter pilots into combat with Luftwaffe veterans until their training was complete.[57] The commander of American B-17s in Great Britain, Ira Eaker, wrote to his superiors in August 1942 that he was "now thoroughly convinced . . . that in the future successful bomber operations can be conducted beyond the range of fighter protection,"[58] but this optimism was not inconsistent with his desire to get as much fighter protection as he could. Eaker went on record in a conversation with Air Corps Intelligence the same month that "we would like to have P-38s with us to help us get in so that fighters won't work on us while we are bombing." And, indeed, the first P-38 escort missions were flown on 26 September 1942.

But the escort program came to an abrupt end in October 1942 as a result of the invasion of North Africa. Four fully trained fighter groups, including 412 P-38s, were transferred from England to North Africa over the objections of the Eighth Air Force commanders, effectively depriving the B-17s in England of the American fighters with the longest range.[59] There followed a hiatus of approximately six months before the newest fighter, the P-47, became operational in England. The P-47 was superior to the P-38, but numbers lagged behind requirements while production problems and defects were worked out. The commander of the Eighth Fighter

[56]Cited in Stephen L. McFarland, *Journal of Strategic Studies* 10 (June 1987), "The Evolution of the American Strategic Fighter in Europe, 1942–44," pp. 189–190.

[57]Craven and Cate, *Army Air Forces*, 2:112, 2:231.

[58]Boylan, "Long Range Escort Fighter," p. 86.

[59]Ibid., pp. 87, 97–98.

Command, Major General Frank Hunter, who had the responsibility for protecting the B-17s in the Eighth Bomber Command, stated that he would need twenty P-47 fighter groups in England by August 1943 in order to provide proper escorts for the anticipated bombing missions. Only three groups were actually in place by that date because of production delays. Eaker wrote to Arnold in January 1943 that he would be relying on P-38s to escort his bombers on into France, asserting that this would reduce his losses "by more than 50 percent." "As soon as we get the additional fighter groups," he added, "we will rush the business of getting them into the fight." But later that month, the sixty P-38s that Eaker had slowly reaccumulated for the escort mission were also transferred to North Africa.[60] Air Forces officers have asserted that both Eaker and Hunter were lukewarm to the need for escort fighters, and had to be replaced by Arnold before long-range escort fighters could be utilized properly.[61] The truth seems to be that the utility of strategic escorts was appreciated, but that the absence of such fighters was not seen by the Air Forces as a reason not to bomb Germany.

The Air Forces' bomber commanders were clearly interested in obtaining and using as many fighter escorts as they could long before American bombers began to take heavy losses over Europe. But even if they had obtained the numbers of P-38s and P-47s they asked for, a problem would have remained. Unmodified P-38s and P-47s lacked the range to fly deep into Germany. As long as American bombers flew no further than France, this was not a problem. But deep penetration missions were another story. Did the American Air Forces take adequate steps to create the new capability to fly these longer range escort missions? Could it have learned faster and better how to adapt to wartime conditions to create long-range escorts sooner?

The question of fuel drop tanks for American fighters in Europe appears to be a classic example of losing a battle for want of a nail. Simple, unpressurized external fuel tanks fashioned from steel or even paper extended range and could be dropped when enemy fighters were encountered so that they would not slow the aircraft down in air-to-air combat. Such tanks, manufactured in the thousands, eventually solved the challenge of the German fighters. P-38s were used with drop tanks from the beginning of the American war in the Pacific, first seeing action in November 1942 over

[60]Ibid., p. 110.
[61]McFarland, "Evolution of American Strategic Fighter," pp. 193–94.

Guadalcanal.[62] The American bombers attacking Schweinfurt were escorted part of the way by P-47s with drop tanks, but the tanks were too small to permit the P-47s to fly all the way to the target. German fighters simply waited for the P-47s to turn and go home before attacking the B-17s. Even the best and longest range fighter of the war, the P-51, needed drop tanks to escort bombers over Germany. Drop tanks of adequate size were the key. If they were in use in the Pacific in 1942, if long-range escorts were recognized early in the war as necessary by the Air Forces' leadership, why were appropriate drop tanks not available in adequate numbers in Europe in 1943?

The enemy is an active force that reacts, but not always in the most likely ways, not always, even, in the ways most advantageous to himself. In the case of the American bombing of Germany, the enemy fighter command reacted in ways that made drop tanks unexpectedly useful. Drop tanks *had* been procured by the U.S. Air Forces for use in Europe as early as the spring of 1941, but for non-combat ferry missions. Drop tanks were regarded as unsuitable for combat missions because even if they were self-sealing they were dangerously vulnerable if enemy fighters were encountered. The air combat conditions that prevailed in Europe until 1943 ensured that any American fighters carrying drop tanks would encounter enemy fighters soon after reaching the European continent. The Germans initially had their fighter defenses very far forward, in a thin belt along the Dutch and French coastlines. During 1942, German fighters were deployed in Luftflotte 3 in France and the Netherlands. In 1943, additional fighters were stationed in those countries, and a defensive belt 125 miles back from the coast was established.[63] American fighters would barely have time to become airborne over the English channel before they would have to drop their external tanks, and their ranges would not be extended at all. As one Air Forces general noted in December 1942, if drop tanks were adopted, "we will go in with our belly tanks; we will drop them at the coast because some guy might shoot us at any time after we reach the coast. To my mind, that defeats the purpose."[64]

Retrospective analysis tended to confirm the wisdom of this judgment. The deputy commander of the United States Strategic Air Force in Europe, General Fred Anderson, noted after the war

[62]Boylan, "Long Range Escort Fighter," p. 293.
[63]McFarland, "Evolution of American Strategic Fighter," pp. 191–92.
[64]Boylan, "Long Range Escort Fighter," pp. 85, 95.

that Germany *could* have made the American use of drop tanks in Europe in 1944 impossible by sending up German fighters to challenge the escort fighters over the English Channel. By engaging them in combat far forward, the Germans would have forced the escort fighters to drop their fuel tanks at the beginning of their flights, depriving them of the necessary additional range. In Anderson's words: "Having that stripping then [*sic*] they could move back in and meet the undefended bombers over the target area. . . . [But] they gave us access to the deeper targets before we had to strip the tanks. These were gifts from Goering and his staff."[65] Long-range escort fighters with the necessary maneuverability were only possible through the use of drop tanks, and in 1943 there were sound combat realities arguing against the use of drop tanks.

The rationality of Army Air Forces behavior is confirmed by its actions elsewhere. Certain fighter missions involved operations in areas where early fighter opposition was not likely. In those cases, drop tanks were adopted quite quickly. In the Pacific war, for example, fighters could fly escort missions with drop tanks over long distances between islands. The chances were small that they would meet Japanese fighters early in their flights, when the American aircraft were far from Japanese air bases. So the utility of drop tanks was recognized and the technology developed here long before this was done in the European theater.[66] Similarly, the use of drop tanks for use by American fighters defending the continental United States made sense, because enemy fighters were not expected close to home, and production of tanks for this mission was initiated in September 1941. This production was given a low priority for the reason that no enemy air force yet had bombers that could reach the continental United States from existing or anticipated bases.[67]

Drop tanks were finally introduced in combat in Europe at the end of July 1943. Since this predated the first raid on Schweinfurt, the innovation could not have taken place because of the shock of combat losses there, as several authors have argued. Did earlier combat losses shock the Air Forces into action, or had the German air force changed its fighter tactics to create an opportunity for the use of the tanks? The answer is mixed. Heavy losses did lead to a call for drop tanks from combat commanders. This first request occurred after the 13 May 1943 bomber raid on the German submarine

[65]Ibid., p. 285.
[66]Ibid., p. 289.
[67]Ibid., p. 67.

yards at St. Nazaire. The attacking bombers experienced heavy losses, which prompted the commander of the Eighth Bomber Command, Ira Eaker, to ask for auxiliary fuel tanks for his fighters. The next month, American bombers began to operate during the day against German cities such as Kiel, Hamburg, and Kassel, which were beyond the range of available escorts, and they took losses measuring between 10 and 20 percent. With these losses, urgent requests were issued by Eaker and Arnold for longer range fighters. Most famous is Arnold's 28 June 1943 order to Major General Barney Giles, the assistant chief of staff of the Air Staff for Operations, Commitments, and Requirements: "About six months remain before deep penetration of Germany begins. Within this six months you have got to get a fighter that can protect our bombers. Whether you use an existing type or have to start from scratch is your problem. Get to work because by January '44, I want an escort for all our bombers from U.K. into Germany."[68]

This wartime innovation does appear to have been motivated by the shock of combat losses and not by learning about the enemy. There does not appear to have been a contemporary understanding that enemy tactics had changed so as to make drop tanks more useable than previously believed. Paradoxically, the hardware for drop tanks had existed before the war but had been rejected because it seemed inappropriate given observations of the enemy. Initial wartime learning, in other words, *inhibited* innovation. Indeed, wartime learning about changes in German fighter operations could not have generated the innovation because it was the innovation itself that forced the change in German tactics that made the use of drop tanks possible. Drop tanks worked because German fighters were pulled back from France and the Netherlands in order to protect central Germany. This withdrawal was not a gift from Goering, who opposed the shift. It was the result of the professional advice of the Luftwaffe senior officers *after* American bombers began bombing raids deep into Germany. According to Adolf Galland, the commander of German fighter forces, the strategy of using German fighters far forward for a peripheral defense of Europe was Goering's idea. That strategy, if it had been adhered to, would have prevented the effective American use of drop tanks. Galland and his Luftwaffe colleagues succeeded in reversing Goering's strategy: "I . . . thought this kind of peripheral defense basically wrong. . . . *After* the American daylight raids of Reich territory became a reality,

Goering . . . believed that he could counter them by establishing defensive strips on the outer limits of the German defense area. This concept reflected the political-propagandist wish to keep enemy aircraft as far as possible out of sight of the German population, but in order to do this fighter forces should have been much stronger."[69] The professional military maxim enjoining the concentration of force led the Luftwaffe officers to react to the first American bombing raids on Germany by concentrating their fighters over Germany. Peripheral defenses had spread German fighters too thinly and had resulted in unsatisfactory low kill rates. By the end of July 1943, P-47s with drop tanks were flying with American bombers part of the way into Germany, and their appearance in German skies forced German antibomber fighters to operate from bases in Germany. The withdrawal of German fighters from France began in the fall of 1943, and all German defensive fighters were placed under the control of one ground controller in Berlin in February 1944.[70]

An innovation implemented despite what was known of enemy countermeasures led to changes in enemy behavior that made the innovation successful. Although the innovation was driven by combat losses, it cannot be said to have been a case of organizational learning in the simple sense of learning by observing what did and did not work in the ongoing war. Nor was it a matter of organizational learning in the sense we have been using that term, in which an organization learns how to learn by defining a new measure of strategic effectiveness that links military operations to a strategic goal to provide a new focus that permits the evaluation of alternative modes of fighting. It comes closer to military reform, in which preexisting norms are not met by units in combat, creating a performance shortfall that is addressed by existing organizational routines. Twenty percent losses were clearly outside the norms of acceptance, and drop tanks were part of the organizational repertoire of the Air Forces.

But this characterization misses an important element of wartime innovation, which is desperation. The bombing of Germany had to be carried out if the Allied plan for weakening Germany's military, and in particular, its air force, was to succeed before the invasion of France. Anything that might help the Army Air Forces carry out that bombing would be used. Drop tanks were adopted, not because

[69]Galland, *First and Last*, p. 209.
[70]Ibid., pp. 210, 248, 251, 253; Boylan, "Long Range Escort Fighter," p. 209; McFarland, "Evolution of American Strategic Fighter," pp. 192, 202.

there was an understanding that they stood a better chance of solving the problem, but because something had to be done. Drop tanks were available, they might help, and they were worth a try. What is clear is that the senior leadership of the Army Air Forces was going to carry out the bombing of Germany whether the drop tanks helped or not. Arnold issued his "do it or else" order for long-range escorts but also gave orders that the mass bombing of German aircraft factories would begin in February 1944, whether long-range escorts were available or not, and he accepted the possibility that those raids would suffer twenty percent loss rates, just as earlier unescorted raids into Germany had.[71] The increased size of the fleet of heavy bombers available to the United States in 1944 made this brute-force solution possible. Arnold could have no confidence that the innovation he had ordered would be available. Aviation technology told him that true long-range fighters were impossible, and combat experience told him that drop tanks would not work. He innovated in desperation because there was nothing to lose. And he got lucky.

CONCLUSIONS

Drawing general conclusions about wartime innovation from the cases of the tank, targeting for strategic bombing, long-range fighters, and submarine warfare against merchant shipping is more difficult than generalizing about peacetime innovation. The development of new strategic measures of effectiveness was of clear importance in obtaining resources for the tank in World War I. The development of new strategic measures of effectiveness was essential for the assessment of the new form of submarine warfare practiced by the United States, but was not of major importance in the initiation or development of the innovation. Developing new strategic measures of effectiveness was at the heart of learning how to *perform* the new function of strategic targeting, but was not a part of initiating the innovation. The long-range escort fighter was an innovation born of desperation, not analysis, that was based on a prewar understanding of the importance of escorts and a gamble that the enemy would not take the obvious countermeasures. New strategic measures of effectiveness played a significant role in three of these innovations but were decisive in only one.

[71]Craven and Cate, *Army Air Forces*, 3:33, 3:35.

As to which structural conditions favored wartime development of new strategic measures of effectiveness, no firm conclusions can be drawn. New measures of effectiveness were developed by the British in World War I when intelligence about the enemy military was bad, by the United States Navy in the Pacific when the intelligence was good but badly distributed, and by the United States Air Forces in World War II when military intelligence was good but irrelevant to the strategic problem. Centralized and decentralized commands both had problems using innovations most effectively in the context of wars that lasted about four years. The structural conditions related to intelligence and command relations do not seem to be decisive in explaining the initiation or effectiveness of wartime innovations.

Some limited but useful generalizations about wartime innovation, however, can be drawn. The most obvious is perhaps the most important. Learning from wartime experience how to perform an entirely new military function was in all cases extremely difficult. The cases studied in chapters 4 through 6 represent relatively successful examples of new capabilities being created after conflict began, and in each case, the success was only relative and imperfect. Truly large-scale production and strategically effective use of the tank did not occur until the last months of the war. The strategic targeting function generated an effective target set only in the last year of the war. Long-range escort fighters with drop tanks that could accompany bombers along their entire route were available only during the last four months of the air war in Europe. The officer corps that commanded U.S. submarine warfare against Japanese merchant shipping was successfully transformed by mid-1943, but the strategic impact of the innovation was not understood until the war was over.

Compare these wartime innovations with the cases of peacetime innovation. American carrier aviation was able to fight successfully from the start of the war and won a decisive victory barely seven months into hostilities. The first ever midocean amphibious assault against a fortified island was the successful operation at Tarawa, which validated the concept and equipment that would be used, with some adjustments for the rest of the war. Helicopter aviation won its first battles in Vietnam, and if the war was lost, it was not because of flaws in the airmobile concept.

Peacetime military innovation appears to have been more successful in dealing with changes in the character of warfare than wartime innovation. The explanation of the process of wartime innovation laid out in the previous chapter suggests why. Wartime innovation can proceed from wartime learning, but learning that goes beyond improvements in the application of existing organizational routines involves the development of a new measure of strategic effectiveness. The military has to learn what to learn about, and how to learn. If pursuit of the old performance goals only makes the problem worse, then a new strategic goal has to be defined. The way in which military operations can bring the nation closer to that goal has to be worked out, and ways to collect and analyze information that make possible the evaluation of alternative methods have to be implemented and exercised. By definition, a wartime innovation, a new way of fighting created in wartime, will not find the existing military organized to help answer these new questions, so the infrastructure necessary for the intelligence collection and analysis will have to be created. In the case of strategic bomber targeting in Europe and submarine warfare in the Pacific, defining and answering the right questions required an understanding of the enemy economy and the ability to collect economic intelligence in wartime. That capability had not existed before the war and was imperfectly cobbled during it. Finally, once the question has been defined and an answer generated, action has to be taken on the basis of this answer.

All this takes time, and time is short during war. Not as much time is needed to overcome the kind of organizational resistance normally found in peacetime. Initial action was taken quickly in all four of the cases examined. The process of learning a new way of fighting *began* quickly enough, but simply took time to complete. The payoff came late in the war and was of limited value. The need for time will make even the most ardent supporter of innovation skeptical about innovation in wartime. Solutions that come after the enemy has triumphed are of no value at all. Even Winston Churchill, no foe to innovation, fought against new naval programs once war began because they would produce new ships only after two or three years had gone by. In October 1939 he wrote that "it is far more important to have some ships to fight with . . . delivered to date, than to squander effort upon remote construction which has no relation to our dangers." Even the program for ships that would be ready in 1941 would have to "jog along as fast as it can, but the ships we need to win the war must be in commission in 1940."[72]

---

[72]Cited in Martin Gilbert, *Finest Hour: Winston Churchill, 1939–1941* (London: Heinemann, 1983), pp. 18, 58.

Wartime innovation can take place and has taken place without a rigorous analysis of strategic options. Drop tanks for long-range fighters were introduced and the submarine force transformed before the analysis was done, but if the account of those two cases presented in this chapter is accurate, the drop tanks succeeded because of a lucky break, and strategic advantage of the campaign against Japanese merchant shipping was never taken.

Peacetime military innovations, in contrast, have decades to work themselves out. The supporting analytical infrastructure can be developed before the war begins, so that errors can be quickly recognized and fixed and the strategic impact of the innovation assessed. In the case of carrier aviation, for example, early in the war there were already available well established procedures in the U.S. Navy that enabled it to evaluate and critique the performance of carrier battle groups and to develop modifications of basic concepts of operations.[73] In peacetime, there is time to think through the problems of new ways of fighting so that in wartime mistakes can be easily recognized and established organizational routines employed to effect reforms.

Added to the intellectual and organizational problems that delay wartime learning and innovation are the problems of collecting intelligence. Even with Ultra intelligence, serious gaps remained, and plain mistakes were made in the analysis of the bombing of Germany. Taken together, the problems of wartime innovation make it much less attractive than it initially seems. In peacetime, the military has to hazard guesses about the character of future war. In wartime, the opportunity to learn from experience does exist. But the empirical record of innovation in peace and war and the hypotheses about the necessary components of wartime innovation suggest that peacetime innovation ought to be pursued, because wartime innovation is so terribly difficult.

[73]See, for example, the circular memorandum with attachments evaluating the multicarrier task force and its ability to handle air attacks in CINCPAC File No. Pac-95-jb, A4–3, Commander in Chief U.S. Pacific Fleet to Commander Air Force, Pacific Fleet, "Operations of Carrier Task Forces," 21 March 1943, Operational Archives, Washington Naval Yard, Washington, D.C. See also Admiral King's bulletins, which he used to develop and promulgate lessons learned about carrier and other naval operations in "Secret Information Bulletin #1: Battle Experience from Pearl Harbor to Midway," 15 February 1943, United States Fleet, Headquarters of the Commander in Chief, Operational Archives, Washington Naval Yard, Washington, D.C., pp. 3:9, 5:6.

TECHNOLOGICAL INNOVATION

[7]

What Is the
Enemy Building?

Military innovation has been thus far defined in terms of major changes in the behavior of military organizations, changes in how they fight or organize for war. This definition has deliberately excluded the question of technological innovation, the process by which new weapons and military systems are created, except insofar as new technologies can be identified as the source of organizational change. Since the focus has been on organizational change, nontechnological developments—amphibious warfare, counterinsurgency, and strategic targeting for aerial bombing, for example—have been considered as much innovations as the development of carrier aviation, escort fighters, and helicopter warfare. In this chapter and the next, the particular problems of purely technological innovation in the military will be examined. In the United States, technological innovation is the business of the military research and devlopment (R&D) communities. Although those communities perform many functions other than developing conceptually novel weapons, the term R&D will often be used as shorthand for "military technological innovation."

Chapter 1 laid out two sets of problems associated with technological innovation in the military. First, it was noted that theories of arms races and economic analyses of military research and development both assumed that technological innovation in the military took place in the context of good information about enemy military technology. It will be the essential argument of this chapter, however, that in major cases involving the United States Army and Air Force in the period 1930–1955, the organizations responsible for research and development did not, in fact, have good information

[185]

about German and then Soviet military technology. A wartime case involving the development of electronics warfare in Britain will also be examined, to provide a contrasting example where good intelligence about enemy technology was available.

Chapter 1 also noted that was difficult to find any other objective way of identifying desirable new military technologies than by taking account of developments among potential enemies. Thus American planners had to choose from among development alternatives without objective knowledge of what would ultimately prove most useful. Was it possible for planners to devise strategies for coping with this radical uncertainty? The question will be addressed in chapter 8.

INTELLIGENCE AND TECHNOLOGICAL
INNOVATION BEFORE WORLD WAR II

How do scientists and engineers working in research and development learn about foreign technology? In practice, such information comes through many channels, formal and informal. Aeronautical engineers observe new craft at air shows and have informal discussions with their foreign counterparts. Engineers and scientists in other fields exchange information informally at conferences. Scientific and technical journals provide access to unclassified developments. Unclassified information can often provide insights regarding matters that remain secret, as for example, in the case of Paul Rosbaud, the scientific advisor to the German publishing house Springer Verlag in the 1930s. From his vantage point he had access to German scientists doing research in nuclear physics and in light, nonferrous metals, used for aircraft construction. A persuasive case has been made that he was able to use his network of scientific contacts as the basis for a private espionage war against the Nazis, an effort that delivered to the British the famous 1939 "Oslo Report" on German research and development in radar, guided missiles, and the proximity fuse.[1]

Unfortunately, the flow of informal intelligence into the research and development establishments of the military is difficult to reconstruct, while official intelligence reports remain archives where they can eventually be studied. As a result, it will be assumed here that

[1]Arnold Kramish, *The Griffin: The Greatest Untold Espionage Story of World War II* (London: Macmillan, 1986), pp. 17–18, 50, 61, 66–67.

the bulk of the information on classified foreign technology came to the attention of American scientists and engineers through normal intelligence channels. This assumption is supported by the testimony of scientists and engineers who confirm that in the cases studied they were ignorant of foreign military technologies except insofar as they were made aware of them through formal intelligence reports.

The declassification of military records in the United States now makes possible a study of what intelligence was available to the U.S. military research and development community from the 1930s through the middle of the 1950s. What the records reveal is a pattern of organizational behavior in which the community, both in peacetime and in war, was surprisingly divorced from intelligence about enemy technology. This resulted from difficulties both in collecting intelligence and in circulating it within the defense establishment so that it reached the R&D community. This changed sometime around the middle of the 1950s, but many of the key decisions about new land warfare systems in the interwar period and about the first generation of long-range offensive nuclear weapons appear to have been made with surprisingly little intelligence input.

From the end of World War I through the 1930s, the Ordnance Department of the U.S. Army was responsible for the development of new weapons and equipment for the field forces. As the official history of that department acknowledged, "as war is competitive and military equipment satisfactory only if it is as good as or better than that of potential enemies," intelligence concerning foreign technology was essential to the proper functioning of the Ordnance Department.[2] An advisory group composed of civilian and military ordnance specialists, the Technical Staff, was established; part of its charter was "to keep informed of the trend and progress of ordnance development at home and abroad."[3] Despite the formal awareness of the need for such intelligence, the Ordnance Department's knowledge of even the most basic trends in foreign military technology was poor. The department was almost entirely dependent on the General Staff's military attachés, who had no technical background and who were not competent to collect or evaluate technical intelligence. The Ordnance Department was able to identify officers for attaché duty who had engineering backgrounds and

---

[2]Constance McLaughlin Green, Harry C. Thomson, Peter C. Roots, *The Technical Services and the Ordnance Department: Planning Munitions for War* (Washington, D.C.: GPO, 1955), p. 208.
[3]Ibid., p. 33.

knowledge of a foreign language. In this era, however, military attachés were required to meet the costs of diplomatic service out of their own pockets. This sum was not trivial, amounting in London, for example, to an additional $10,000 a year over army pay, at a time when middle-class salaries were a fraction of that sum. As a result, during the entire period 1920–1940, the Ordnance Department was able to send only nine technically qualified attachés abroad; during the entire period November 1930–May 1940, there were exactly two technically qualified Ordnance attachés in Europe, one in Moscow from 1934 to 1940, and one in London 1936 to 1940. One attaché was able to obtain useful information about German antitank guns and the international trend to heavier antitank guns, which the U.S. Army was unknowingly ignoring. His reports had to be forwarded through the regular army attaché in Berlin, who did not endorse his findings, and consequently they seem to have carried no weight back in the War Department.[4]

The consequences of this intelligence failure were serious. The U.S. lacked a competitive tank engine in the period before World War II, but in the absence of hard data to the contrary, the chief of the Ordnance Department was able to assert in 1938 that U.S. tank engines were the best in the world.[5] The lack of an antitank weapon competitive with European weapons had its roots in the desire of army combat commanders to keep their divisions light and tactically mobile; heavy antitank guns and correspondingly heavy armor were successfully resisted until 1944. The small caliber antitank weapons in the U.S. arsenal were justified on the grounds that they could penetrate the armor on U.S. tanks. No data on the armor or penetrating capabilities of foreign tanks and weapons entered into the debate.[6] The judgment of the army's official history on prewar technical intelligence is harsh, stating that whatever intelligence "filtered through to the [Ordnance] Department was casual and tended to leave research to proceed in a near vacuum." A postwar review of what information the War Department did have in its files on German military equipment showed "how much misinformation the reports contained." General George Marshall testified in 1947 that before the war U.S. military intelligence data was confined to what its attachés "could learn at dinner, more or less over the coffee cups."[7]

[4]Ibid., pp. 208–11.
[5]Ibid., p. 203.
[6]Ibid., pp. 184–85, 280–86.
[7]Ibid., page 260–261. Marshall quoted in hearings before the U.S. Senate, 80th Congress, 1st session, 30 April 1947.

In contrast, when Germany began its clandestine program to rearm, it opened intelligence channels that would enable it to keep tabs on foreign military technology. The Krupp industrial combine established relations with the Swedish Bofors artillery company in 1921. Until 1935, Krupp gave Bofors full access to Krupp patents in return for the access to the Bofors factories and "technical information on current developments."[8]

Nor did matters within the U.S. Army improve much in the European theater during World War II. Liaison between the British scientific intelligence establishment and the American army was good in the area of electronics warfare against Germany. Where the U.S. Army was not able to rely on British intelligence, the situation was abysmal. The War Department's history of the Scientific Branch of the Military Intelligence Service of the War Department General Staff is blunt: there was no central army intelligence organization assigned to scientific intelligence at all until early 1943. Until then, the tasks and information had been scattered through the Signals Corps, the intelligence office of the Army Air Staff, the Protective Security Branch of the Signals Corps, the Office of the Air Communications Officer, various consultants to the secretary of war, the Office of Strategic Services (OSS), and the civilian Office of Scientific Research and Development (OSRD). Not surprisingly, it was found to be impossible to pull together from these various agencies a coherent picture of enemy radio and radar operations from which to consider the development of countermeasures. Eventually a central office was created, but it was limited to three officers who were not given access to Ultra intelligence until March 1945. This office had no long-term research background or files. It had no technically trained translators, and "a great deal of material was lost in translation because the assistance of technically trained personnel was not available to the translators who consequently were unable to recognize certain material as potentially valuable." The efforts of the Scientific Branch to collect scientific intelligence of use to the field commands "was made difficult by 'insulation' between the field and the Branch. It is believed that liaison resulting from visits to combat areas would have resulted in more timely and more useful intelligence adapted to the express requirements of the operational commands." The branch personnel were qualified engineers, but there were no scientists on the staff, and it was candidly admitted that the "quality of the reports of the Scientific Branch would have been

[8]Ibid., p. 248.

improved if a small number of pure scientists (as contrasted with technicians) had been available." Obtaining the necessary scientific expertise through consultation with outside scientists in the OSRD was sometimes possible but was often blocked by security requirements. The list of possible sources of technical intelligence that were left unexploited by the branch "included foreign technical publications, interviews with scientists and technicians having experience in foreign countries, patent applications, and pre-war industrial agreements for interchange of technical information."[9]

The records of the other sections of the Military Intelligence Service of the War Department General Staff, in particular the Research Unit, the Air Subsection, and the Air Industry Section, do not appear to have concerned themselves with technical or scientific intelligence at all, having been entirely devoted to the collection of enemy order of battle and aircraft production.[10]

In the case of the U.S. Army before and during World War II, it is not possible to observe the impact of intelligence on military R&D, since little intelligence was made available to the weapons designers. There were, however, a few wartime cases in which intelligence was available and was utilized, and these can be examined to determine the impact the information had on the development of new equipment.

INTELLIGENCE AND THE INVENTION OF ELECTRONICS WARFARE

Electronics warfare is one of the most significant military technological innovations of the modern era. Research and development was begun in many countries in the interwar period and continued after the outbreak of World War II. The Office of Scientific Research and Development (OSRD) in the United States will be discussed more fully in the chapter on scientists and military innovation. Here, it is sufficient to note that in World War II that office spent a total of $457 million on contract research to support the military from the time of its creation by President Roosevelt in June 1941 un-

9The material in this paragraph is drawn from "History of the Intelligence Group, Military Intelligence Service, War Department General Staff (WDGS), Scientific Branch," SRH-130, Record Group 457, National Archives, Washington, D.C., pp. 010–019, 027.
10"History of the Intelligence Group, Military Intelligence Group, WDGS, Military Branch," parts I, IV, and VI, SRH-131, Record Group 457, National Archives, Washington, D.C.

til the termination of its operations in November 1945. Of the total, $128 million was spent on radar and radar countermeasures, more than on any other category of research. Much of that money was for the creation of actual operational systems that had not existed even in prototype.[11]

Electronics warfare is a field in which intelligence is particularly important to innovation. The connection between a military unit that broadcasts or receives radio waves and the countermeasures that seek to neutralize such broadcast or reception has always been obvious. Initial U.S. Navy resistance to the radio as an instrument of ship-to-shore communication was based on the observed difficulty of preventing other transmissions from interfering with or jamming naval communications.[12] The first use of radio for commercial news also marked the first use of radio jamming. Results of the America's Cup yacht races off Newport Rhode Island in September 1901 were transmitted by a reporter using the Marconi radio network. When his transmission was over, the reporter left the radio on to broadcast a continuous tone, "jamming" any potential rivals who might also try to transmit the news by radio. U.S. Navy exercises simulating the defense of Maine against an invasion fleet in 1903 used radio jammers against the notional enemy.[13] The first use of jamming in wartime occurred in the Russo-Japanese War, when the Russian military jammed the transmissions of Japanese sailors who were spotting the fall of naval gunfire against Port Arthur in order to adjust its aim.[14]

In peacetime, new weapons took years to develop, and years were available for the development of countermeasures. The intelligence collection process could be adjusted to the pace of technological development. Wartime conditions, however, created a cycle in which intelligence gained from battlefield operations was quickly incorporated into new equipment and practices. This made intelligence collection a much more urgent matter. In World War II, the American

[11]Irwin Stewart, *Organizing Scientific Research for War: The Administrative History of the Office of Scientific Research and Development* (Boston: Little, Brown, 1948), pp. 92, 232.

[12]Susan J. Douglas, "Technological Innovation and Organizational Change; The Navy's Adoption of Radio, 1899–1919," in Merritt Roe Smith, ed., *Military Enterprise and Technological Change: Perspectives on the American Experience* (Cambridge: MIT Press, 1985), p. 128.

[13]Louis A. Gephard, *Evolution of Naval Radio-Electronics and Contributions of the Naval Research Laboratory* (Washington, D.C.: GPO, 1979), p. 299.

[14]Alfred Price, *The History of U.S. Electronics Warfare: The Years of Innovation—Beginnings to 1946* (Westford, Mass.: Association of Old Crows, 1984), pp. 3–6.

Countermeasures Committee, a subcommittee of the Joint U.S. Communications Board reporting to the Joint Chiefs of Staff, noted: "Countermeasure requirements are determined by operational considerations and it is impossible to plan long in advance. . . . Techniques change rapidly and equipment must be available on very short notice in order to be of value."[15] By operational considerations the committee meant the current electronics practices of the enemy and the observable "research and development trends which are indicated for future radar and radio sets."[16] Having identified the need for intelligence to direct the development of countermeasures, the committee noted its dissatisfaction with the lack of information about enemy research and development and expressed a desire to know where the enemy research labs were and on what frequencies the Germans were operating.[17]

After the war the U.S. agency responsible for developing new electronics countermeasures summarized the principles it had developed for directing research and development. The standard operating procedure had been established. First, "the operating characteristics, location, and tactical use of the enemy equipment must be ascertained," after which jamming and deception equipment can be designed to take advantage of the vulnerabilities of enemy weapons. Frequent intelligence updates and organizational agility are needed to insure that radio countermeasures remain effective, because the enemy "is calling the tune" and could shift frequencies, the location of transmitters, and operating procedures. Detailed, specific electronics intelligence was judged to be the prerequisite for the success of the research and development establishment.[18] The director of the Telecommunications Research Establishment (TRE), which developed the new weapons of electronics warfare for Great Britain, made the same point most succinctly: "Before TRE could produce suitable jammers, it had to know exactly what it was to jam."[19]

---

[15]Countermeasures Committee, Joint Chiefs of Staff (JCS) Decimal No. 334, folder 8/27/42, Record Group 218, National Archives, Washington, D.C., minutes of meeting 23 December 1942.

[16]Ibid., minutes of meeting 3 February 1943.

[17]Ibid., minutes of meeting 12 May 1943.

[18]C. G. Suits, *Summary Technical Report of Division 15, National Defense Research Council, Radio Countermeasures* (Washington, D.C.: U.S. Office of Scientific Research and Development, 1946), pp. 1, 9–10.

[19]Martin Streetly, *Confound and Destroy: 100 Group and the Bomber Support Campaign* (London: Jane's Publishing, 1985), p. 15.

Despite its recognized importance, however, electronics intelligence in all three major western powers, the United States, Germany, and Great Britain, was, for differing reasons, poor before World War II. The United States Naval Research Laboratory began work on radio controlled aircraft and flying bombs soon after the end of World War I, and research in this field led quickly to an interest in jamming the radio controls of any such weapons the enemy might have. Frequency scanners were developed to search for enemy transmitters, and by the early 1930s working models were ready.[20] But in the United States the opportunity and incentive to use these devices in the field against potential European enemies was limited by foreign policy. Dr. Luis Alvarez, one of the first scientists working on countermeasures to radar at the Radiation Laboratory at MIT (a laboratory for military radar research under the auspices of the National Defense Research Committee) found that in the period 1940–1941 the United States had essentially no information on enemy development in radar. To rectify this inadequacy he instituted a program to build receivers to search out enemy radar transmissions. Even after the United States entered the war, however, there was a serious intelligence vacuum. The director of the radar countermeasures program, Dr. Frank Terman, recalled the intelligence situation in March 1942: "We really did not know whether one needed 0.5 watts, 5 watts, or 500 watts to jam a radar. Nor did we have much information of the enemy equipment. We knew that the Germans had radar, but we had few details. We had no idea at all whether the Japanese had even heard of radar."[21]

Prewar British intelligence about German electronics weapons was limited for technical reasons. German radar used the Very High Frequency band, which gave better performance, but which, given the technology of the time, had limited range, so that radar testing in Germany was not detectable in Great Britain. In the fall of 1939, after the war had begun, the British military attaché in Oslo did receive intelligence from an unknown source about German scientific military developments, information which came to be known as the Oslo report. Among the technical capabilities disclosed was radar for early warning of approaching aircraft, which had detected British

[20]Price, *The History of U.S. Electronics Warfare*, p. 7.

[21]Ibid., pp. 14, 23. The electronics countermeasures division of the Radiation Lab was later separated from the lab for security reasons, becoming Division 15 of the Office of Scientific Research and Development, with the title of "Radio Coordination."

bombers attacking Wilhelmshaven in September 1939 at a range of 120 kilometers. The report, however, was not credible to the British military service intelligence branches. Even photographs of radar antennae on the wreck of the German pocket battleship *Graf Spee* taken in June 1940 were insufficient to convince British military intelligence of the existence of German radar.[22]

The Germans, on the other hand, made at least one serious effort to determine the existence and nature of British radar. In the summer of 1939 they flew specially equipped Zeppelins in the area of the Bawdsey Station research laboratory run by Sir Robert Watson-Watt, the father of British radar. Strong lower frequency signals were detected. Because for radar such lower frequency transmissions were inferior to the Very High Frequency signals used by the Germans, and because the signals had some of the same characteristics as those that might come from the British electrical power grid, the Germans dismissed the possibility that the signals represented British radar transmissions. In fact, the British radio industry was not capable of producing Very High Frequency transmissions with the necessary power and so was forced to use the lower frequency signals for radar.[23]

Intelligence to support the development of electronics warfare did improve in Great Britain and the United States once the war began in earnest, and the effort produced impressive benefits. By tracking wartime intelligence as it related to the development of electronic countermeasures (ECM) and by then examining the estimates of the impact of ECM on combat operations, some sense of the usefulness of intelligence in guiding the development of new technologies can be obtained.

Initial efforts to establish an institutional mechanism within the British defense community that would provide intelligence to the designers and operators of new military equipment encountered considerable resistance. R. V. Jones sought in December 1939 to establish a central Scientific Intelligence Section. His initiative at first

---

[22]R. V. Jones, *The Wizard War: British Scientific Intelligence 1939–1945* (New York: Coward, McCann, and Geohegan, 1978), pp. 69, 93. R. V. Jones stated before his death that he had been able to identify the man who supplied the Oslo report and that it was not the man identified by Arnold Kramish in *The Griffin* (London: Macmillan, 1987). Jones refused to identify the man he claimed was the source of the report, however, on the grounds that the man was still alive.

[23]Jack Nissen and A. W. Cockerill, *Winning the Radar War, 1939–1945* (New York: St. Martin's, 1987), pp. 28, 33–38. Nissen worked in Watson-Watt's laboratory and conducted interviews after the war with German radar scientists that uncovered the story of the German use of Zeppelins for electronics intelligence work.

failed because of Royal Navy opposition to efforts that might affect its own research work, but was then given the necessary impetus by the German bombing of Great Britain in 1940. Interrogation of German bomber crews shot down over Great Britain supplied evidence that some form of radio beam navigation system was in use, but the scientific wisdom in the private, commercial research world was that it was impossible for radio beams to "bend" sufficiently to cover the area over Great Britain from transmitters in Germany. Provoked by a comment by one German POW that the English would never find the special equipment, Jones carefully searched the wrecks of German bombers and discovered that radio receivers used for landing at night or in poor weather were far more sensitive than they needed to be for this purpose. With this information and additional POW interrogation reports, Jones in July 1940 was able to persuade Churchill's scientific advisor, Frederick Lindemann, and then Churchill himself to devote the necessary manpower and equipment to a search for the radio transmissions used by the Germans. Since invasion of the United Kingdom was expected in three weeks, only such manpower as would not affect war production could be spared. Five radar sites along the coast were dedicated to the search. The result is well known: the detection of the beams and the creation of a new section under Robert Cockburn at TRE for the construction of electronic jammers and deception devices.[24]

What is noteworthy is that despite the brilliant success of this combined intelligence and development effort, no immediate efforts were made to expand the collection of intelligence concerning German radar. The German army had photographed the large British radar towers across the English Channel as soon as they conquered France,[25] and British intelligence had picked up suggestions that the Germans had installed radar on the French coast that had assisted in the sinking of British ships, but the German installations were too small to show up in normal British photoreconnaissance, and doubts about the existence of German radar continued. Thus when the Royal Air Force, before the planned Bomber Command offensive, asked Jones in February 1941 to study German night defenses, there was no formal program for listening to German radar transmissions. Jones had to assemble an ad hoc group for this mission. From this intelligence and from Ultra intercepts that gave the num-

---

[24]Jones, *The Wizard War*, pp. 74–75, 94–103, 127. See also Alfred Price, *Instruments of Darkness: The History of Electronic Warfare* (New York: Scribner's, 1978), p. 32.
[25]Nissen, *Winning the Radar War*, p. 88.

ber of German radar sets used for the defense of the coastlines of Bulgaria and Romania, Jones successfully inferred the operating characteristics of the German Freya and Wurzburg radars. The Bruneval raid that successfully seized and returned to England a Wurzburg radar confirmed his inferences and provided information on German radar production rates. Combined with a captured map that showed the locations of antiaircraft searchlight installations in Belgium, Jones was able to piece together a picture of the operating characteristics of the German air defense line, the so-called Kammhuber Line, and to develop offensive tactics and jamming equipment that could be used to attack its vulnerabilities.[26] In the case of the Kammhuber Line the vulnerability proved to be operational rather than technical; it resulted from the division of the line into nonoverlapping sectors, any one of which could be overwhelmed by "streaming" large numbers of bombers through it and overloading its capacity to handle enemy aircraft.

British intelligence efforts continued in an ad hoc fashion that nevertheless on occasion yielded impressive results. The raid against Dieppe in August 1942 had as one of its objectives the capture of a Freya early warning radar. The mission failed in this objective, but the technical personnel attached to the raid did succeed in cutting the telephone lines connecting the Freya to other German commands. This forced the radar station to broadcast its data instead of sending it over the telephone lines. The British were able to intercept its broadcasts and from the resulting intelligence developed the first successful jammer of the Freya radar, the Mandrel jammer.[27] The Mandrel was put into action in December 1942.[28] A German night-fighter pilot defected with his radar-equipped airplane in 1943, giving the British enough intelligence to design the Serrate system, which enabled the British to locate German night fighters by homing in on their radar transmissions.[29]

One of the most simple and most effective ECM devices during the war was chaff, or "Window." It owed its development both to intelligence and to a theoretical understanding of the way in which properly sized strips of foil could be used to generate strong radar

---

[26]Jones, *The Wizard War*, pp. 122, 189–93, 245, 267–68, 277, 290.
[27]Nissen, *Winning the Radar War*, pp. 192–95. Nissen was the technical operative assigned to the raid. Because of his knowledge of British radar and countermeasures, his military unit at Dieppe had orders to shoot him if it appeared likely that he would be captured by the Germans.
[28]Price, *Instruments of Darkness*, p. 129.
[29]Ibid., p. 165.

echoes. With this technique a small amount of material dropped in front of a formation of bombers could effectively disable the radar of an opponent. When, over North Africa in 1941, a Wellington bomber equipped with special antennae for intelligence collection noticed that it was drawing unusually heavy radar directed antiaircraft fire, an officer had the presence of mind to measure the length of the antennae on his aircraft. He discovered that the rods measured half the length of the radio waves used by the German radar, creating a resonant effect that strengthened radar reception. This fit well with the theoretical understanding of radio waves and led to the first use of chaff, over North Africa in 1941 (with no noticeable results).[30] The first major use in the European theater was during the 24–25 July Bomber Command raid on Hamburg, in which British losses were reduced to 1.5 percent, one-fourth the previous average of 6 percent. The Germans quickly recovered by switching to fighter tactics that did not rely on radar, and within one month, by the time of the 23 August 1943 Bomber Command raid on Berlin, Bomber Command losses were back up, to 8 percent.[31] The Germans continued to work on countermeasures; within three months after the Hamburg raid, three antichaff modifications had been added to 60 percent of the Wurzburg radars used to direct fighters and antiaircraft artillery.[32] The introduction of the chaff by no means ended the need for new equipment and the intelligence to guide its development.

British electronics intelligence collection was not put on a firm organizational footing until December 1943. Although RAF units were formed to collect electronics intelligence from aircraft penetrating German defenses as early as December 1940, the shortage of aircraft and electronics personnel resulted in the units being diverted to other missions. By the end of 1943, however, German success in keeping at least even with British ECM prompted the creation of a RAF bomber group, Group 100, dedicated to electronics intelligence collection and ECM missions. Group 100 integrated into its structure scientists and technicians from the TRE, so that the intelligence collectors, ECM operators, and designers of new equipment could communicate directly.[33]

The results of this combined intelligence–R&D effort are difficult to ascertain. On one level, British ECM was disappointing. Bomber

[30]Streetly, *Confound and Destroy*, pp. 19–20.
[31]Ibid.; Price, *Instruments of Darkness*, pp. 170–71.
[32]Price, *History of U.S. Electronics Warfare*, Appendix E, extract from ABL-15 report, "Intelligence Information on RCM Effectiveness in the ETO," 16 June 1945.
[33]Streetly, *Confound and Destroy*, pp. 31–43.

Command losses remained high throughout the war. But once the war ended, the British had an unusual opportunity to test the effectiveness of their ECM efforts. In 1945 they captured intact the German air defense command for Denmark, and decided to conduct Operation Post Mortem, during the last week of June and first week of July 1945. Operation Post Mortem was a full-scale, mock Bomber Command raid against the German system, complete with the German wartime air defense staff, to evaluate the effectiveness of Bomber Command ECM. There were many artificialities in the operation. Obviously, no German fighters were allowed to fly. Beyond that, the Danish system had not been attacked during the war, so its operators were not as experienced as those in other German commands. Still, the lessons were striking. First, British assumptions as to how the German system worked were found to be totally accurate. Second, the use of chaff caused the Germans to overestimate the size of the raid by a factor of ten. Third, while German equipment did incorporate some useful antichaff features, those same devices were unusually vulnerable to jamming, so that the effectiveness of the combination of chaff and jamming was considerable.[34]

What this had meant in operational terms was harder to measure. How much more effective would German defenses have been in the absence of ECM? The United States government found that interrogations of the crews of German antiaircraft guns, the commandant of the German school for training those crews, and German radar scientists all estimated that ECM reduced German antiaircraft effectiveness 70 to 75 percent. This figure was roughly confirmed by the fact that a good German 88-millimeter antiaircraft gun crew around Berlin used an average of eight hundred shells to shoot down one aircraft before the heavy use of ECM, and about three thousand shells during the period in which ECM was in heavy use.[35]

ECM was a set of new technologies that did not exist before the war. Its development was, after some initial organizational delays, closely tied to intelligence and thus gives some indication of the effectiveness of a research and development program that is guided by intelligence as a matter of routine. The commander of the German ECM establishment, General Martini, by contrast, argued that his effort to counter the British was hindered by the shortage of person-

---

[34]Ibid., pp. 113–24; Price, *History of U.S. Electronics Warfare*, pp. 195, 287.
[35]Price, *History of U.S. Electronics Warfare*, Appendix E.

nel and other resources during the war, but also by his inability to fly intelligence collection missions against Great Britain after 1940.[36]

It must be emphasized that the intelligence relevant to technological innovation was not only that which concerned the technical parameters of enemy equipment. An understanding of the enemy's organizational habits also led to the identification of vulnerabilities. The German penchant for automated systems created particular vulnerabilities, since it was easier to fool an automated system than an intelligent human operator.[37] The United States found that certain very sophisticated automatic jamming systems carried on U.S. bombers performed extremely well and could help compensate for the lack of trained operators but were themselves vulnerable to agile German operators who learned how to spoof the systems into shutting down. Conversely, the Japanese proved to be relatively immune to deception efforts, not because they were particularly sophisticated, but because their system was so rigid that it tended not to react to anything, including deceptive practices.[38] It was necessary to understand not only how the enemy machinery worked but also how his human organization functioned if effective new technologies were to be deployed.

Electronics warfare in World War II provides a relatively clear case of the utility of intelligence about the technical capabilities and operating practices of the enemy as a guide for weapons innovation. Enemy weapons and practices could be observed in use or perhaps detected in development a few months before they were introduced in combat. While such intelligence collection is conceivable in peacetime, it was not performed in the 1930s. Wartime differed from peacetime in one major respect. Measures that could neutralize enemy weapons had to be introduced quickly, within a matter of months, if they were to have an impact either before the war ended or before the enemy introduced a new and different weapons system. If the opportunity to observe the enemy's equipment was greater in wartime, the time available in which to do so was extremely limited.

Peacetime conditions differ markedly. Without the urgency of war, there are much longer periods of development for weapons. However, the question how to respond to the enemy becomes more com-

---

[36]Price, *Instruments of Darkness*, p.187.
[37]Ibid., p. 49. Robert Cockburn commented of one German system that by making it automatic, the Germans made its defeat "a piece of cake."
[38]C. G. Suits, *Radio Countermeasures*, p. 350.

plex. The relevant question is not what capabilities the enemy has today, but what weapons he is likely to have five or ten years from now. If the United States were to make use of intelligence for technological military innovation in the postwar world, it would have to not only develop or retain the capability to track Soviet technological developments but also acquire something very new, the ability to make projections about what the enemy was likely to be doing over the next ten years. The capabilities of collecting intelligence and making projections were both slow to develop in the postwar Western intelligence community.

THE TECHNICAL INTELLIGENCE RESPONSE TO THE COLD WAR

During World War II, the United States developed electronics intelligence collection capabilities against Japan but remained dependent on Great Britain for data in the European theater. Without any technical intelligence capabilities of its own in Europe, and with the general demobilization of the U.S. intelligence apparatus after the war, the fate of British intelligence was of considerable relevance to the continuing ability of the United States to incorporate scientific and technical intelligence into its research and development process in the immediate postwar period. The irony is that the successful program of Allied electronics intelligence and technological innovation came to a halt at war's end, with consequences for the development of major weapons systems in the postwar world.

Along with the rest of the West's military establishment, the electronics intelligence collection branches of the United States and Great Britain largely shut down in 1945. A few B-29s equipped with electronics intelligence packages continued to fly out of Alaska in 1946 and 1947,[39] but the British electronics warfare establishment disbanded in December 1945, on the assumption that the Soviet Union was backward electronically and the atomic bomb had in any case made electronics warfare in support of bombing missions unnecessary. A Radio Warfare Establishment was created but kept on a minimal budget until the early 1950s.[40] This deemphasis is consistent with the general disestablishment of the wartime scientific intelligence apparatus in Great Britain portrayed by Peter Wright in his memoirs of his years as a scientist in the British Security Service.

[39]Price, *History of U.S. Electronics Warfare*, p. 257.
[40]Streetly, *Confound and Destroy*, p. 131.

Wright asserts that he was approached in 1949 by both the Security Service and the Secret Intelligence Service when in the wake of the Berlin crisis the British government was considering "how best to galvanize the scientific community once again." In that interview, the situation was outlined for Wright by the chief scientist of the Ministry of Defense: "It had become virtually impossible to run agents successfully behind the Iron Curtain, and there was a serious lack of intelligence about the intentions of the Soviet Union and her allies. Technical and scientific initiatives were needed to fill the gap." The organizational response in Great Britain appears to have been slow. R. V. Jones, despite his wartime accomplishments in this area, is said by Wright to have been rejected by the British intelligence establishment. In words attributed to one senior Security Service officer, "Jones has been making a play for the job [of Chief Scientist], but if we let him in, he'll be wanting to run the place the next day."[41] Data were in shortest supply from 1945 to 1951. Studies of the British signal intelligence establishment suggest that the defection of Soviet diplomats and cypher clerks and the initiation of aircraft flights along Soviet borders to listen in on signal traffic, combined with poor Soviet communications security, had begun to improve intelligence collection by 1951.[42]

For its part, the United States was putting considerable effort into intelligence collection in Europe in the aftermath of the war, but its focus was on the analysis of German economic and technological developments during the war. This work began very soon after that country was occupied and had a major impact on U.S. weapons development programs. This survey of German programs produced some of the first intelligence projections in the area of military technology. But this was quite different from a sustained effort directed at the new military competitor of the United States, the Soviet Union.

As the war against Germany was proceeding toward its conclusion, the chief of staff of the U.S. Army Air Forces, General Hap Arnold, asked Theodore von Karman, the director of the Guggenheim Aeronautical Laboratory at the California Institute of Technology (later the Jet Propulsion Laboratory), to survey current scientific and technological developments and project likely trends as a guide to Air Force weapons development. Arnold told von Karman, with

[41]Peter Wright, *Spycatcher* (Richmond, Australia: William Heinemann, 1987), pp. 5–6, 25.
[42]Andy Thomas, "British Signals Intelligence after the Second World War," *Intelligence and National Security* 3 (October 1988), 103–10.

whom he had worked on technical Air Force issues since 1936, to forget the technologies of World War II: "What I am interested in is what will be the shape of air power, in five years, or ten, or sixty-five." Von Karman was given a free hand, and access to all the countries allied with or under the military control of the United States. He assembled a team of scientists and arrived in the United Kingdom in April 1945. Over three million German scientific documents were at his disposal, and numerous interviews with German scientists, including Walter Dornberger and Werner von Braun, were conducted. In August, his report, *Where We Stand*, was submitted to the Air Force.[43] It related that while German radar development had, with some exceptions, lagged behind that of the United States and Great Britain, Germany had devoted more resources than the Allies to supersonic aerodynamic testing and modeling and had developed "arrowhead" designs for wings that could reach and surpass the speed of sound without becoming unstable.[44]

This information had immediate and dramatic effects on weapons innovation in the United States. Design work on the next generation of U.S. bombers, to replace the B-29 and B-36, had begun in response to a statement of military requirements in November 1944, and the Boeing aircraft company had developed plans for the first American jet bomber by December of that year, a design with four turbojet engines and a straight wing. A Boeing engineer, however, was part of the team von Karman took with him to Germany, and his knowledge of German aeronautic advances led immediately to the complete redesign of the Boeing bomber. This design became the B-47, the first swept-wing bomber.[45]

More generally, on the basis of its survey, the report made a set of projections that "for future planning of research and development . . . [were] to be considered as fundamental realities." These included the possibility of supersonic flight, unmanned aerodynamic systems capable of delivering weapons payloads at ranges of up to several thousands of miles, target seeking antiaircraft

[43]Michael Gorn, *Harnessing the Genie: Science and Technology Forecasting for the Air Force, 1944–1986* (Washington, D.C.: GOP, 1988), pp. 12–14, 21–26.

[44]Theodore von Karman, *Where We Stand: A Report Prepared for the AAF Scientific Advisory Group*, May 1946, Headquarters Air Material Command, Wright Field, Dayton, Ohio, available on microfilm at the Air Force Historical Division, Bolling Air Force Base, Va., pp. 1–2, 49.

[45]Thomas Marschak, Thomas Glennan, Jr., and Robert Summers, *Strategy for R&D: Studies in the Microeconomics of Development: A Rand Corporation Research Study* (New York: Springer-Verlag, 1967), p. 126.

missiles, the need for supersonic offensive systems to penetrate the new antiaircraft systems, systems for perfect communication between fighters and ground control stations, and all-weather navigation systems.[46]

These projections, impressive in their prescience, were based on the study of German laboratories. The von Karman team did visit the Soviet Union but was given essentially no access to facilities. The von Karman projections represented an assessment of what was technically possible on the basis of a survey of some of the most advanced facilities in the world. They did not provide a direct look at what the *Soviet Union* was actually doing, whether its technical progress was along the same lines of as those of Germany, or whether its technical development and operational practices would resemble those projected by von Karman. Since the Soviet Union was known at that time to be making vigorous efforts to recruit, kidnap, or otherwise obtain the services of German military scientists, the state of German research and development provided some insights into what the Soviet Union might be doing. The Joint Intelligence Committee of the U.S. Joint Chiefs of Staff warned at the end of 1945: "Unless the migration of important German scientists and technicians into the Soviet zone is immediately stopped, we believe that the Soviet Union within a relatively short time may equal the United States developments in the fields of atomic research and guided missiles and may be ahead of U.S. development in other fields of great military importance, including infra red, television, and jet propulsion."[47] The United States thus had some knowledge of potential Soviet military research and development.

This inferential intelligence, however, was different from direct hard intelligence, collected and analyzed, about the Soviet military technology. From the first five to ten years after World War II there are numerous indications that such intelligence was in short supply. The Research and Development Board in the Office of the Secretary of Defense, the heir to the wartime Joint New Weapons and Equipment committee of the JCS, prepared in September 1948 its "Technological Estimates of New Weapons and Countermeasures," in which it reported that U.S. electronics warfare developments were being confined to the preparation of a few pieces of equipment for training purposes. Weapons development in this area was limited

[46]Von Karman, *Where We Stand*, p. iv.
[47]Cited from JCS records, National Archives, in Tom Bower, *The Paperclip Conspiracy: The Battle for the Spoils and Secrets of Nazi Germany* (London: Michael Joseph, 1987), pp. 206, 226–27, 291.

because not enough was known about Soviet radars and military electronics. The report's survey of foreign technologies noted that neither was anything known of Soviet antiaircraft weapons development, beyond the assumed availability to Soviet scientists of basic techniques and knowledge in the radar field.[48] The Guided Missiles Committee of the Research and Development Board periodically prepared "Technical Estimates" of developments in its field. These reports are characterized by broad assumptions about the nature of Soviet missile work. The first report, prepared in September 1947, assumed that the Soviets were working on bringing captured German missile technology, in particular, German surface-to-air missile prototypes, to production and were trying to improve the performance of German V-2 rocket engines. Subsequent Technical Estimates did not even try to make any assessments or projections of Soviet missile work.[49] A separate committee of the Research and Development Board assembled a list of general questions about Soviet military technology in August 1947. Did the Soviets have ENIAC or IBM type computers to support missile systems? How had they modified the U.S. radars they had received during World War II under the Lend Lease program? Did they network their air defense radars? Where were their missile test ranges, and what kind of instrumentation did those ranges have? Were the Soviets sticking with the German use of alcohol and liquid oxygen for rocket fuel or were they experimenting with new fuels? In short, the whole range of data about the rate and direction of Soviet missile technology which was requested by the committee responsible for supervising the development of the major new weapons system of that period, guided missiles, had yet to be collected. In December 1950, the committee revisited the questions and noted the persistent lack of data about Soviet programs.[50]

American war plans for the period 1945–1950 also displayed a lack of information about Soviet guided missile programs and air defense systems. War plans are different from operational military plans and do not always contain detailed intelligence information about enemy

[48]EL 80/5, September 1948, Committee on Electronics, Research and Development Board, Office of the Secretary of Defense, Record Group 330, National Archives, Washington, D.C.

[49]GM 28, Technical Estimates, 29 September 1947, June 1949, 30 March 1951, Guided Missiles Committee, Research and Development Board, Record Group 330, Office of the Secretary of Defense, National Archives, Washington, D.C.

[50]GM 29, Technical Intelligence Regarding Foreign Guided Missiles, Research and Development Board, reports of 27 August 1947 and 20 December 1950, Record Group 330, Office of the Secretary of Defense, National Archives, Washington, D.C.

targets. However, American war planners needed to know the number and general location of Soviet strategic weapons. They assumed that the guided weapons used by the Germans in World War II would be available to the Soviets. Joint Intelligence Committee reports of 1946 used in preparing war plans anticipated that the Soviet Union would have large numbers of V-1 and V-2 type weapons with nonnuclear warheads by 1949. The "Broadview" survey of the air defense requirements of the continental United States undertaken in 1946 assumed that after 1950 the Soviet Union would have long-range missiles capable of striking at the United States, and the Joint War Plans Committee estimated in 1947 that the Soviet Union had the capacity to launch 80 V-1 and 25 V-2 type attacks against Great Britain a day in order to neutralize that country as a forward base for American strategic bombers. The assumed use of V-1 and V-2 type weapons against Great Britain was retained in the revised war plan "Broiler" in early 1948.[51] But these estimates and warplans contained no apparent information about the number or even the general location of V-1 and V-2 type missile launch sites, whether they were fixed or mobile, protected or not.

The war plans seem to reflect an absence of precise knowledge of the location of key military targets.[52] It was an open question whether it was possible to destroy such targets with an American attack. The Joint Intelligence Committee noted in 1948 that target data was poor, and so even if a twenty-kiloton nuclear weapons was successfully delivered against a Soviet city, the research and development centers and other military targets assumed to be in the environs of that city might not be destroyed.[53] This lack of data could be partly compensated for by intelligent guesswork. Defense analysts later pointed out that it could safely be assumed that air bases from which Soviet heavy bombers could be launched would be those that had been built up through World War II, the locations of which were known from German Air Force intelligence.[54]

[51]Joint Intelligence Committee 375/1, 29 November 1946; Joint Staff Planners 24 October 1946, Broadview Survey; Joint War Plan Committee study 474/1, 15 May 1947; Joint Strategic Planning Group revised plan Broiler, 11 February 1948; all cited in Steven T. Ross, *American War Plans 1945–1950* (New York: Garland, 1988), pp. 10, 35, 41, 65.

[52]Ross, *American War Plans*, p. 14.

[53]Ibid., pp. 14–15, 83.

[54]Interview with Andrew W. Marshall, 2 August 1985, conducted by James Digby and Joan Goldhammer, Rand Corporation, 1985, p. 28. Andrew Marshall was on the professional staff of Rand from 1949 through 1972 and participated in several major target studies during the 1950s.

Soviet missile bases, however, did not have long organizational histories and could not be located. This intelligence gap had implications for American weapons development. Could guided missiles be developed for attacks against the Soviet missile bases? This depended on how far away those bases were. If they were located near the borders of the Soviet Union, shorter-range missiles based in Turkey or Germany might be very effective against them. If Soviet missile bases were located in the center of the country, longer-range missiles would be needed. If Soviet launch sites were few in number and not widely dispersed, they might be vulnerable to a barrage of inaccurate missiles with twenty-kiloton weapons. If they were dispersed and protected, first-generation guided weapons might be ineffective against them. Data was also needed to determine whether guided weapons were better or worse than manned bombers for this mission. Knowledge of Soviet air defenses was necessary to evaluate the relative capabilities of new manned bombers, cruise missiles, and ballistic missiles to penetrate to their targets.

Just when this kind of information became available in the United States is still unclear. A picture of the intelligence uncertainties that prevailed in the United States in 1950 can be reconstructed from the first major effort to evaluate the ability of the nation's strategic air forces to execute the war plan against the Soviet Union. The debates in the United States over the desirability of the B-36 bomber as opposed to other weapons systems, in particular the "super carriers" of the *America* class proposed by the navy, and the general problem of managing interservice disputes over new strategic weapons technologies prompted Vannevar Bush, the preeminent defense scientist of the period, to suggest to Secretary of Defense James Forrestal that he create a new, neutral agency to do objective evaluations of American weapons systems. Forrestal created the Weapons System Evaluation Group (WSEG) in 1948, responsible both to the secretary of defense's Research and Development Board and the Joint Chiefs of Staff.[55] WSEG's first task was to evaluate the capability of the United States to execute its air war plan against the Soviet Union. WSEG was given full access to JCS war plans and intelligence sources. It submitted its report in February 1950.[56] Among other findings, it

---

[55]John Ponturo, *Analytical Support for the Joint Chiefs of Staff: The WSEG Experience, 1948–1976* IDA S-507 (Arlington, Va.: Institute for Defense Analysis, July 1979), pp. 22–34.

[56]JCS 1952/11, Weapons System Evaluation Group Report No. 1, Record Group 218, JCS Decimal No. 373, 10-23-48, Sections 4–6, National Archives, Washington, D.C. (hereinafter WSEG #1).

cast doubt on whether the B-36 would be better at penetrating Soviet air defenses than the B-29 and B-50 it was to replace.[57]

For the purposes of this discussion, however, the most important finding of the study was that intelligence about the Soviet Union necessary for weapons system evaluation was extremely poor, particularly in the area of ECM: "U.S. knowledge of Soviet capabilities in the field of electronic countermeasures [which included early warning radar, fighter ground control systems, and jamming] is essentially nil and U.S. readiness in this respect is not demonstrated. Should one side or the other achieve marked superiority in this field, large improvements or impairments in effectiveness might result."[58] While some information about Soviet radar sites along the country's perimeter was available, the report concluded that "our present knowledge of the electronic weapon structure of air defense in the USSR is based on a tenuous argument derived from fragmentary and often contradictory bits and pieces filtering in to us from satellite countries, Germany, Austria, and derived from the knowledge of equipment available to the USSR through Lend Lease, through German scientists and captured equipment, and through the British and American technical press." The report further noted that there existed in 1950 nothing like the "Y" intercept branch of British wartime intelligence that had been created to supply data to the U.S. and Great Britain about enemy radar and radio systems. It concluded that "the present deficiencies in intelligence . . . seriously limit SAC's [Strategic Air Command's] capabilities to execute a successful ECM campaign.[59]

Intelligence uncertainties extended beyond electronics warfare to air defense order of battle. How many fighter airplanes would SAC bombers encounter as they penetrated Soviet air space? Soviet fighter units had been identified and counted, but how many fighter aircraft were actually available? How many of the regiments were "hollow," with few operationally ready aircraft or with no aircraft at all? In the words of the report, "Sufficient data is not available to permit an accurate count of the number of fighter aircraft assigned to the majority of air regiments."[60]

The first director of research of WSEG and the man who supervised the preparation of WSEG Report 1 was Philip Morse. A phys-

[57]Steven L. Rearden, *History of the Office of the Secretary of Defense: The Formative Years, 1947–1950* (Washington, D.C.: GPO, 1984), pp. 409–10.
[58]WSEG #1, pp. 161–62.
[59]WSEG #1, Enclosure D, pp. D40–D41.
[60]WSEG #1, Enclosure C, p. C7.

icist specializing in wave behavior, Morse had been involved in the development of antisubmarine weapons and tactics in World War II in his work for the National Defense Research Council. During that period he realized the need for reliable, first-hand data on the performance of weapons systems in the field. Morse used POW interrogations and other sources to develop new weapons and tactics to counter German acoustic homing torpedoes.[61] Given this background, he was acutely sensitive to the near intelligence vacuum about the Soviet Union in which WSEG was operating in 1950. He later recalled that WSEG had to

> estimate as well as . . . [it] could what defenses a possible enemy (the Russians, for example) might have a year or so in the future. We had to extrapolate from World War II data. . . . The biggest uncertainties came in the estimates of Russian defense capabilities [based on] . . . spies, chance observations of Russian equipment, comments from occasional defectors, or obscure paragraphs in Russian papers. All these fragments were put together by people with whom we had no direct contact and were "filtered" by still others, using procedures carefully concealed from us. . . . [T]he reports we did get did not produce confidence. Estimates of numbers and of effectiveness would vary widely from month to month, sometimes by factors greater than two.[62]

Because of its ignorance concerning Soviet radar networks, jamming capabilities, number of fighters, and the nature of the Soviet fighter control system, the WSEG assessment of SAC's overall ability to penetrate Soviet air defenses had to be based on assumptions, rather than knowledge. To hedge against uncertainties, WSEG assumed two different levels of Soviet air defense capabilities. In one model, WSEG assumed that Soviet air defense organization was poor, and lacked effective early warning, good ground control for fighters, and the ability to conduct nighttime intercepts. In the other, WSEG assumed that the Soviet air defense system had the best features of both the British and German wartime air defense systems.[63] Against this superior air defense, WSEG estimated that SAC bombers would succeed in releasing nuclear weapons against 80 percent of the 220 assigned targets, but would lose one-third of their forces in doing so. These losses would preclude SAC from at-

---

[61]Philip Morse, *In the Beginning: A Physicist's Life* (Cambridge: MIT Press, 1977), pp. 180–82, 203.
[62]Ibid., pp. 252–54.
[63]WSEG #1, Enclosure E, p. E3.

tacking with nonnuclear weapons the targets that in the "Offtackle" warplan would have to be destroyed or reattacked once the U.S. stockpile of nuclear weapons had been used up.[64] It is important to note that in 1950, nuclear weapons were still scarce commodities, and it was assumed that conventional bombing of the Soviet Union would have to be conducted once nuclear weapons were expended. In warplan "Dropshot" of 19 December 1949, for example, the use of 180 atomic bombs against the Soviet Union and 73 bombs against East Europe in the first thirty days of the war was planned. Thereafter, 38,000 tons of nonnuclear bombs would be dropped on the Soviet Union and its allies in East Europe. A B-29 carried five tons; thus the plan called for successful nonnuclear sorties, clearly impossible with anything like the attrition rates computed by WSEG.[65]

In the absence of hard intelligence about Soviet air defense capabilities, it was more difficult to determine exactly what kind of offensive weapons to develop. The WSEG report, perhaps reflecting this, concentrated less on the evaluation of alternative weapons systems and more on the ability of the U.S. military to meet its own internal planning goals. In particular, SAC was measured in terms of its logistical ability to support its planned level of bomber missions. This kind of evaluation required no data about the enemy, but simply a knowledge of how many air bases on the periphery of the Soviet Union SAC B-29s would need in order to reach their targets, and how much aviation gas would have to be stockpiled at those bases. In terms of this internal planning standard, WSEG judged SAC to be unable to execute its warplan.[66]

In short, in 1950 the Weapon System Evaluation Group found itself severely constrained in its ability to evaluate new weapons technology because of a lack of intelligence data. This lack was reflected in the two recommendations. The first was that SAC rethink its warplan, given its inability to provide the necessary logistical support. The second was that "grave deficiencies" in intelligence be corrected in order to "improve the basis for future planning and evaluation."[67] The impartial group established by the secretary of defense to advise him on alternative military technologies reported that it could not do so because of inadequate intelligence.

A similar problem faced American officials who were trying to evaluate what defensive systems needed to be developed to defend

[64]WSEG #1, p. 160.
[65]Ross, *American War Plans*, pp. 125, 127–28.
[66]WSEG #1, pp. 158–59.
[67]WSEG #1, p. 161.

against Soviet nuclear bombers. The Central Intelligence Agency submitted a special estimate to President Truman in October 1951 that evaluated the Soviet military capability to attack the United States in 1952. While it is possible that sources of intelligence were available that could not be used in this report, the copy designated for the President's own use is available and is likely to provide a reasonably comprehensive picture of U.S. intelligence about Soviet weapons systems. Like WSEG Report 1, it reveals as much what the United States did not know as what it did. The report projected that the Soviet Union would have one hundred seventy-kiloton weapons and approximately one thousand B-29 type bombers, the TU-4, by 1952.[68] Beyond that, the report could say very little. How well would the Soviet bombers perform? What tactics would they use? Would they come over in massed formation, dispersed in small units, or even on individual missions? Would they fly at night or only in daylight? Answers to these questions would have major implications for the kind of air defenses the United States would have to develop. The report was candid about U.S. ignorance: "Very little information is available on the tactical doctrine of Soviet Long Range Aviation. It is considered that the Soviets would attempt to deliver the maximum possible weight of attack within the shortest possible period of time." They might or might not attack at night or in bad weather. They might or might not have navigation good enough to allow them to attack individual military facilities or factories, as opposed to large centers of population.[69] Soviet use of radar for navigation at night and electronic countermeasures to thwart U.S. air defenses were unknown, with the last hard information being that concerning the World War II equipment they had obtained from the United States or seized from the Germans.[70] From which airfields would Soviet bombers be launched? Three areas in the Soviet Union provided bases that were within TU-4 range of the United States, and the report noted all the runways in those areas that were known in 1945 to have been capable of handling medium bombers. It then warned that "discussion of these three areas is based on information which is far from adequate." In connection with air bases in the Kola Peninsula it offered this further disclaimer: "Very little information is available on the present status of these airfields.

---

[68]"Soviet Capabilities for a Military Attack on the United States Before July 1952," SE-14 [hereinafter SE-14], Central Intelligence Agency, 23 October 1951, available from Harry S Truman Library, Independence, Mo., pp. 1, 3.

[69]SE-14, p. 3.

[70]SE-14, pp. 6–7.

Some may have been improved to accommodate medium bombers. Scattered and unconfirmed reports of base improvements have been received."[71] The question of what new defensive systems or offensive weapons for counterattack were needed was left very much open. Studies by the Rand Corporation in 1951 on possible new designs for air defense systems encountered similar problems with a lack of intelligence. The character of U.S. air defense systems depended in part on how effective U.S. attacks on the Soviet Union would be. But there was little intelligence available to Rand on the capabilities of Soviet air defenses, and analysts responded by assuming that the Soviet air defense engineers would use the same equipment and practices available to their U.S. counterparts.[72]

After 1951 it becomes more difficult to determine what intelligence was available to American officials responsible for the development of new weapons. Gregg Herken reports that by 1952 Air Force intelligence officers and those civilians with contacts with Air Force intelligence had precise knowledge not only of targets identifiable from German war intelligence but also of other strategic target locations. Intelligence from intercepted radio communications had provided this data, and technical intelligence sources had provided information about the capabilities of Soviet strategic bombers.[73] Studies of British signals intelligence in this period suggest that it improved after 1948. Soviet cypher clerks defected, providing the West with information that could be used to break codes. Soviet security discipline was not as strict as it was later to become. And the RAF began flying more intelligence missions, using bombers equipped with radio receivers to fly over Germany to Berlin and then around the southern borders of the Soviet Union.[74]

But what did the United States government actually know? Any attempt to use declassified U.S. government documents to determine the extent of its knowledge of the Soviet military is made more difficult because the declassification of government documents to this point may well have favored reports which did not utilize the most sensitive intelligence data. Documents which remain classified may well reveal that the United States government had more information than the declassified record revealed in 1989. Enough infor-

---

[71]SE-14, pp. 3–4.

[72]30 March 1989 interview with James Digby by Stephen P. Rosen. Digby was one of the principal authors of the Rand Air Defense Study, R-227, 15 October 1951.

[73]Gregg Herken, *Counsels of War*, expanded ed. (New York: Oxford University Press, 1987), pp. 80–81.

[74]Thomas, "British Signals Intelligence after the Second World War," pp. 104–7.

mation has been made public, however, to make possible a rough sketch of what data was available, not necessarily at the most classified levels, but to those individuals working directly on research and development.

Intelligence about Soviet radar and radios appears to have developed unevenly in the 1950s. While deficiencies remained in some areas, progress was clearly made in others. In mid-1952, the Joint Chiefs of Staff were still urgently querying the British about their knowledge of the radar and radio frequencies used by the Soviet Union, presumably to help the U.S. begin its own electronics listening program. In November 1952, the U.S. Defense Department agency responsible for intercepting and decrypting radio signals and electronics intelligence, the National Security Agency (NSA), was created out of the Armed Forces Security Agency, which, in turn, had been created out of the Army Security Agency.[75] In 1953, the NSA complained that there were inadequate facilities for sharing U.S. and British intelligence.[76] In 1954, both the NSA and the CIA were complaining that the electronics warfare programs of the U.S. military were still inadequate, and it was not until mid-1954 that the Joint Intelligence Committee established bureaucratic routines for reporting electronics intelligence.[77]

At the same time, it is clear that some information about Soviet weapons developments was getting to the United States. Unofficial histories report, for example, that U.S. radar sites began to operate in Turkey in 1954 that were able to track Soviet ballistic missile test flights. It was determined that the Soviets were testing rockets with a range of eight hundred miles, much greater than that of the captured V-2s.[78] Herbert York was a physicist who had worked on the Manhattan Project and the fusion bomb. He was a member of the Air Force Scientific Advisory Board and the "Teapot" committee, under the chairmanship of mathematician John von Neumann, that was appointed to review strategic cruise and ballistic missile programs. In his memoirs, he states that the intelligence available to the Teapot

[75]James Bamford, *The Puzzle Palace* (New York: Penguin, 1983), pp. 15, 64, 71.

[76]Reports on Electronics Warfare Conferences with the United Kingdom, 28 May 1952, 20 July 1953, JCS Decimal No. 337, Record Group 218, Joint Chiefs of Staff, National Archives, Washington, D.C.

[77]Countermeasures and Radio Intelligence and Jamming Equipment, 22 January 1954 and 6 May 1954, JCS Decimal No. 311, Record Group 218, Joint Chiefs of Staff, National Archives, Washington, D.C.

[78]Edmund Beard, *Developing the ICBM: A Study in Bureaucratic Politics* (New York: Columbia University Press, 1976), p. 186.

committee at the beginning of 1954 came primarily from two sources: German scientists who had been taken to work on Soviet missile programs at the end of World War II but who were now returning to the West and Soviet defectors. York singles out the reports of Colonel Grigory Tokaty-Tokaev as being representative of the kind of data supplied by defectors.[79] Tokaty published his own account of Soviet weapons programs in the early 1950s in which he relates that Stalin had personally told him of his desire for ballistic missiles capable of reaching the United States. Stalin asked if Tokaty realized "the tremendous strategic importance of machines of this sort? They could be an effective straitjacket for that noisy shopkeeper Harry Truman." In 1947, Tokaty was placed in charge of project TT-1, an effort to develop a three-stage transatlantic rocket with an initial mission of launching an earth-orbiting satellite sometime between 1950 and 1952. In a separate program, V-2s had been improved; by 1949 a large single-stage rocket with a range of about nine hundred kilometers was in full-scale production, as were V-1 type missiles. Tokaty's ICBM project, however, appears to have become involved in a political struggle: "For reasons having nothing to do with the project itself or our professional qualifications, we found ourselves in a difficult position. Towards the end of 1947 our work was paralyzed. Some of us were compelled to seek refuge in the West, others were arrested; the rest had to wait." It was not until 1954 that the project was restarted; it produced the SS-6, the first Soviet ICBM and space launch vehicle.[80]

The Teapot committee's review of U.S. missile programs was supported by analysts at the Rand Corporation working on technical analyses of potential ICBMs. The leader of the Rand team was Bruno Augenstein, who published his analysis independently in a special Rand Memorandum on 8 February 1954. In his unclassified report, Augenstein strongly hinted that there was intelligence that the Soviets were actively engaged in building their own ICBM. After summarizing the technical virtues of the ICBM, Augenstein wrote: "Our enemies can gain strategic advantages from a similar development and there is good evidence that they are making progress in that direction. In a very real sense, the prosecution of our own bal-

---

[79]Herbert F. York, *Making Weapons, Talking Peace* (New York: Basic Books, 1987), p. 93.
[80]G. A. Tokaty, "Soviet Rocket Technology," in Eugene M. Emme, ed., *The History of Rocket Technology: Essays on Research, Development, and Utility* (Detroit: Wayne State University Press, 1965), pp. 279–82.

listic missile program is a race against time."[81] Augenstein later stated that this was a sanitized version of a highly classified report prepared in December 1953 utilizing the best available intelligence about Soviet rocket work. It had proved impossible to circulate that report because of the sensitivity of the information about Soviet efforts and American nuclear weapons design developments, so the Rand special memorandum was written in which much vaguer language was employed.[82]

The Teapot Committee report itself gives mixed indications about the intelligence available about Soviet activities. Classified top secret, the report is cautious in its estimate of Soviet activities: "The available intelligence data are insufficient to make possible a positive estimate of the progress being made by the Soviet [sic] in the development of intercontinental missiles. Evidence exists of an appreciation of this field on the part of the Soviet and of activity in some important phases of guided missiles which could have as an end objective the development by the Soviet of intercontinental missiles. While the evidence does not justify a conclusion that the Russians are ahead of us, it is also felt by the Committee that this possibility certainly cannot be ruled out." The report noted also the importance of understanding Soviet defenses against manned bombers in analyzing the case for U.S. missiles, but ventured no assessment of those defenses, a silence consistent with the intelligence problems the field outlined earlier. Von Neumann, however, entered his personal judgment that the Soviets would strengthen their air defenses sufficiently by the second half of the 1950s to warrant a U.S. missile program of the greatest urgency and that the Soviet strategic missile programs, while largely unknown, nonetheless warranted "grave concern."[83]

From these sources, it is reasonable to infer that at the beginning of 1954 the Americans responsible for development of strategic offensive weapons had reason to believe that the Soviets were working seriously on ballistic missiles, and that their programs were almost certainly as competent as those in the United States. Beyond that, there was disagreement. Reasonable men with access to classified

[81]B. W. Augenstein, "A Revised Program for Ballistic Missiles of Intercontinental Range," Special Memorandum No. 21 (Santa Monica, Calif.: Rand Corporation, 8 February 1954), p. 1.

[82]Interview with Bruno Augenstein conducted by James Digby and Joan Goldhammer, 22 May 1986, Rand Corporation, 1986, pp. 11–12.

[83]"Recommendations of the Strategic Missile Evaluation Committee" (von Neumann committee) RW008-4, 10 February 1954, available from the Air Force Historical Division, Bolling Air Force Base, Va., pp. 1, 11.

data disagreed about whether the Soviets might be ahead of the United States in this field and about how quickly their air defenses would improve. They knew that Soviet ICBM programs would eventually yield results but very little about development schedules, technical characteristics, or production and basing plans. Intelligence available to the strategic weapons development community had improved, but not in a dramatic fashion.

The perception grew among the scientists and engineers involved in weapons system development that more and better intelligence about the Soviet Union was needed. The Air Force sponsored two study groups in 1953, projects Vista and Beacon, and both highlighted the intelligence problems facing that service. In response, the Air Force's Scientific Advisory Board, a group of distinguished civilian scientists retained to advise on the development of new weapons, assembled an Intelligence Systems Panel under the chairmanship of James G. Baker. Motivated primarily by the need to locate Soviet ICBM "launching sites and supporting installations, [and] to keep these under surveillance and to anticipate actual hostilities by a sufficient margin to prepare our own countermeasures," the Baker panel, renamed the Reconnaissance Panel, called in March 1954 for the development of a high altitude reconnaissance airplane having no other function than to monitor the Soviet Union in peacetime.[84] This recommendation was significant not only for its role in the development of the U-2 but also for its explicit acknowledgment of the continued lack of information about the Soviet ICBM program, the need for tactical early warning of Soviet attack, and the need for intelligence to support weapons development. It is of course possible that data were available within the government that would not be made available to outside scientists. We have seen how even the WSEG, which was cleared for access to warplans, did not have direct access to the most sensitive intelligence data. It would not be legitimate to infer that no additional intelligence data were available anywhere within the government about Soviet ICBM research and development. It is important to note, however, that the Air Force Scientific Advisory Board, a group whose formal responsibility was to think about the future technological requirements of the U.S. Air Force, felt that it did not have adequate intelligence to do its job, whether or not this was because information was withheld.

---

[84]Thomas A. Sturm, *The USAF Scientific Advisory Board: Its First Twenty Years, 1944–1964* (Washington, D.C.: GPO, 1967; available from the USAF Historical Liaison Office), pp. 48–49, 62.

Additional evidence suggests that the U.S. defense and intelligence establishment was slow to respond to the need for better intelligence to support the analysis of new military technologies, and slow to share intelligence when it did become available with analysts outside intelligence and operational commands. Despite the weaknesses noted in the studies above, the Air Force chief of staff for intelligence in 1949, Major General Charles Cabell, maintained that the organization and technical capabilities of the Air Force photo reconnaissance units were adequate. This provoked a member of his staff, Lieutenant Colonel Richard Leghorn, to respond: "I still cannot bring myself exactly to share this view . . . for the relative importance of air reconnaissance as an instrument to collect intelligence about such a system as the Russian is great indeed." The Air Force itself noted in October 1950 that "the present AF holdings of USSR photography are both out of date and extremely incomplete."[85] The problem persisted into 1953, when Leghorn wrote to the chief of staff of the Air Force that "qualitative intelligence and reconnaissance capabilities constitute the primary problems, and without extraordinary action, these might delay adoption at operational planning levels of strategies with an emphasis on counterforce operations."[86]

As this memo indicates, the Air Force in the early 1950s often had to be convinced that it needed better intelligence. The Air Force saw its primary intelligence responsibility as providing target data to its offensive operational forces. This was a consequence of the dominance of the Strategic Air Command over Air Force intelligence, a dominance that led the Air Force Technical Intelligence Command under General Harold Wilson to focus intelligence collection on current military operations rather than on longer term analysis.[87] The problem of intelligence capabilities and priorities was called attention to yet again in early 1955 in a report to the president by a commission chaired by James Killian: "We must find ways to increase the number of hard facts upon which our intelligence estimates are based [as well as to obtain better warning]. . . . To this end we recommend adoption of a vigorous program for the extensive use, in many intelligence procedures, of the most advanced knowledge in

[85]Merton Davies and William Harris, *Rand's Role in the Evolution of Balloon and Satellite Observations and Related U.S. Space Technology* (Santa Monica, Calif.: Rand Corporation, 1988), pp. 19, 22.
[86]Ibid., p. 37.
[87]Interview of Merton Davies, 3 April 1989, by Stephen P. Rosen, Santa Monica.

science and technology." One month later, the U.S. government issued its General Operational Requirement No. 80 for advanced reconnaissance satellites. In addition to photography, a requirement for electronics intelligence collection capabilities was laid out.[88]

Even after better intelligence became available, the problem of analysis and distribution remained. Work on new U.S. weapons and capabilities had to be based on good intelligence about current Soviet and *projected* Soviet capabilities. The Rand Corporation had been established by the Air Force precisely to foster forward-looking analytical capability, but the Air Force had neglected to provide Rand with institutional arrangements that would give it adequate and appropriate intelligence support. Individuals at Rand who did have access to highly classified intelligence because of their particular job experiences noted a serious data problem at Rand. Andrew Marshall, who later became the director of strategic studies at Rand and the director of net assessment in the Office of the Secretary of Defense, and Joseph Loftus, who had come to Rand from Air Force intelligence, prepared a study of U.S. reconnaissance capabilities in April 1954. Marshall recalled the problems of that time:

> Well, both of us were very troubled by what we thought was the poor quality, the inadequacy of intelligence input into the Rand studies which grew out of, I think, two things. One was that while the Air Force would supply us with some of the projections that were made by the intelligence people, the intelligence people themselves had never been forced to do really good 5 to 10 year projections at that point. And so they weren't probably the best sort of thing for Rand studies.
>
> I think Rand was seen by the Air Force as a group of people that were focused on the future, and therefore didn't need to know about intelligence on current targets et cetera. One of the first things that Loftus began talking to me about was how inaccurate he thought the people at Rand's understanding was about what it was we *really* knew about the Soviet Union.[89]

In order to correct this problem, Marshall and Loftus instituted Project Sovoy. Selected Rand analysts would "provide interface with the intelligence community, and provide better forecasts—intelligence forecasts—and also . . . collect through [Loftus's] connections . . . with the intelligence community kinds of material which

[88]Davies and Harris, *Rand's Role*, pp. 61–62.
[89]Interview with Andrew W. Marshall, 2 August 1985, by James Digby and Joan Goldhammer, Rand Corporation, pp. 15–16.

they otherwise wouldn't send to Rand."[90] The impact of Project So-voy was limited, in Marshall's view, to changes in the perceptions of a few Rand analysts of the character of military developments in the Soviet Union.[91] Rand as an institution was less affected, at least in part because it remained difficult for analysts to obtain security clearances that would allow them access to the best U.S. intelli-gence. No one at Rand, for example, had access to the data collected by the U-2 flights, and senior Rand analysts did not have access to various forms of sensitive communications intelligence until the late 1960s.[92] Since Rand was initially founded by the Air Force to do research on "intercontinental warfare, other than surface, with the objective of advising the Army Air Forces on devices and techniques,"[93] the inability to supply Rand analysts, either directly or indirectly, with a systematic flow of the intelligence necessary to analyze Soviet military capabilities and behavior represented a sig-nificant break in the link between enemy activities and technological innovation in the United States in the period to the mid-1950s.

CONCLUSIONS

The impact of military intelligence, both good and bad, on the initiation of quantitative arms competitions has received public at-tention in the United States for several decades. Secretary of De-fense Robert McNamara commented on the role of faulty intelligence in the 1961 decisions leading to the quantitative build up in U.S. strategic nuclear offensive forces.

> In 1961, when I became Secretary of Defense, the Soviet Union pos-sessed a very small operational arsenal of intercontinental missiles. However, they did possess the technological and industrial capacity to enlarge that arsenal very substantially. . . .
> Since we could not be sure of Soviet intentions . . . we had to insure against such an eventuality by undertaking ourselves a major build-up of the Minuteman and Polaris forces.

---

[90]Ibid., p. 20.

[91]Interview with Andrew Marshall, 26 April 1989, by Stephen P. Rosen.

[92]Interview with Merton Davies, 3 April 1989, by Stephen P. Rosen; interview with James Digby, 2 April 1989, by Stephen P. Rosen.

[93]Memorandum of Edward Bowles to Secretary of the Air Force Stuart Symington, 26 November 1946, cited in Bruce L. R. Smith, *The Rand Corporation: A Case Study of a Nonprofit Advisory Corporation* (Cambridge: Harvard University Press, 1966), p. 47.

Thus, in the course of hedging against what was then only a theoretical possible Soviet build-up, we took decisions which have resulted in our current superiority in numbers of warheads and deliverable megatons.

But the blunt fact remains that *if we had more accurate information* about planned Soviet strategic forces, we simply would not have needed to build as large a nuclear arsenal as we have today.

Now let me be absolutely clear. I am not saying that our decision in 1961 was unjustified. I am simply saying that it was necessitated by *a lack of accurate information.*[94]

The role of intelligence about *quantitative* changes in enemy forces has received additional study. Albert Wohlstetter has documented the imperfect intelligence projections about the size of the Soviet strategic offensive forces in American defense planning in the 1960s.[95]

The role of intelligence in stimulating *qualitative* changes in military technology has not yet received this kind of attention. The focus of this chapter has not been the universal role of intelligence in the process that results in the development of new weapons. It has not even been a study of intelligence in the American research and development establishment as a whole. This analysis has been confined to the role of intelligence in three specific cases: the U.S. Army's ground forces weapons planning before and during World War II; the development of electronics warfare by the United Kingdom and the United States during World War II; and the analysis of new strategic weapons systems in the United States in the period 1945 to 1955. In these areas, it cannot be assumed that all relevant documents are available for public study. Great care must be exercised in drawing any conclusions from these three cases. No conclusions, for example, about development of naval or tactical aviation weapons can or should be drawn from this study.

However, two conclusions do appear warranted. The first is that U.S. Army and U.S. Air Force strategic weapons development appears to have taken place in a context in which the research and

---

[94]Speech of Secretary of Defense Robert S. McNamara before United Press International Editors and Publishers, San Francisco, 18 September 1967, reprinted in "Scope, Magnitude, and Implications of the United States Antiballistic Missile Program," Hearings before the Subcommittee on Military Applications of the Joint Committee on Atomic Energy, Congress of the United States, 90th Congress, 1st session, 6–7 November 1967, pp. 107–108; emphasis added.

[95]Albert Wohlstetter, "Is There a Strategic Arms Race?" *Foreign Policy* 15 (Summer 1974), 13–14.

development community did not have access to good intelligence about qualitative technological developments among the actual and potential enemies of the United States. This lack of intelligence reflected the difficulties of data collection and organizational problems of distribution.

In face of these problems, in several significant areas American military development was not shaped by intelligence about the enemy's technology developments. This finding has at least two major corollaries. The first is that qualitative technological improvements in the U.S. military were driven by factors other than clear foreign technological threats. Major elements of the military research and development community, for a period of twenty five years beginning in the 1930s and extending into the early 1950s, simply had little relevant intelligence to guide their programs and choices. In the period after World War II, American officials involved in weapons development repeatedly acknowledged this lack of intelligence and called for redress. Whatever drove U.S. technological innovation in the cases discussed, it was not intelligence assessments of foreign capabilities.

The second concerns the exceptional case of British electronics warfare, in which good intelligence about German technology and operational practices was available to the community charged with the development of new weapons. The British and their American allies were able to create a new technological field of warfare that reduced enemy antiaircraft capabilities by 70 to 75 percent. Many other factors went into the Allied electronics warfare victory—a superior industrial infrastructure and bungled German production decisions. But it is not unreasonable to suggest that some measure of the success in this area of military technological innovation was due to the quality and proper handling of intelligence.

These conclusions, tentative though they may be, suggest additional questions. If intelligence was not the primary source of technological innovation in the American military from 1945 to 1955, what was? If American research and development took place in the context of uncertainty about the enemy, what strategies did the responsible institutions use to cope with that uncertainty? These questions are taken up in the following chapter.

[8]

Strategies for Managing Uncertainty

Technological innovations that produced certain major components of the United States military cannot be understood as resulting from a qualitative arms race. Those involved in decisions about new military technologies for the U.S. Army and Air Force simply do not appear to have had access to good intelligence about the Soviet military technological developments. How, then, were decisions made as to technologies to develop?

Military research and development decisions are made amid great uncertainties. In an ideal world, such decisions would be managed by estimating the future cost of alternative programs and their prospective military values, and then pursuing the program with the best ratio of cost to value. But, as was discussed in chapter one, there are tremendous difficulties in forecasting the real value and costs of weapons development programs. These uncertainties, combined with the empirical difficulty American technology managers had in collecting intelligence on the Soviet Union, meant that research and development strategies in the real world tended to become strategies for managing uncertainties. At least two such strategies are conceivable. One of the most politically important can be called, for want of a better phrase, "let the scientists choose." The theoretical and practical case for and against this strategy will be examined below. It will then be compared with the theoretical and practical arguments for a strategy that concentrates on low-cost hedges against various forms of uncertainty.

Scientists and Managing Innovations in Military Technology

Klauss Knorr and Oskar Morgenstern made in 1965 the most explicit argument for giving scientists the leading role in initiating innovations in military technology. They began by acknowledging that the uncertainties inherent in planning military research and development made it impossible to calculate either a "satisfactory" rate of technical innovation or the direction in which such innovation should proceed. They nonetheless asked whether "the current R&D output [is] satisfactory, or is it stagnating or lacking in proper direction? . . . [D]o we give proper nurture to the institutional complexes on which, over the long run, we must rely for imaginative military R&D?" Their answer was no but it was admitted that they and those of a like mind "have not posed the issue in a way permitting their suspicions to be either confirmed or disproved. But neither have those who do not share those anxieties. . . . The facts are that both sets of responses rest largely on hunch, and that it is not easy to give these hunches proper definition and provide them with a rigorous analysis of observable phenomenon."[1]

In the absence of such rigorous analysis, Knorr and Morgenstern could only suspect that the U.S. government was supporting an unacceptably low rate of military technological innovation. Instead of a logically rigorous research and development strategy, their strategy was to pursue a surrogate goal in place of a "satisfactory" rate of military technological innovation. "Inescapably confronting great uncertainties, there is one recourse, however, which we can hardly afford to neglect. This is to maintain an unremitting search for new military technologies, if only on the assumption that a resourceful or lucky opponent might be capable of making any discovery we are able to discern at the technological frontier." By shifting goals in this way, Knorr and Morgenstern also shifted the balance between scientists and military officers in favor of the scientists. They pointed to a distinction between "invention" and innovation. Invention was the creation of a new idea for a weapon, innovation was the choice of which new ideas to develop. Knorr and Morgenstern argued, vaguely, that the military had done neither very well, while scientists had. "It is generally agreed that ideas for new weapons systems have in the past usually come from the scientists and engineers, and

[1]Klauss Knorr and Oskar Morgenstern, *Science and Defense: Some Critical Thoughts on Military Research and Development*, Policy Memorandum No. 32, Center of International Studies, Woodrow Wilson School of Public and International Affairs, Princeton University, mimeo, 18 February 1965, pp. 3–4.

rarely from the military," and that "there seems to be widespread consensus among qualified observers that a rich technological menu has been offered to the military by the scientists and engineers." Nor had the military redeemed itself by choosing well from that menu. "There is appreciably less certainty that the military (and civilians in the defense establishment) have, on the whole, performed the innovating function very well." The reason for this discrepancy was clear. Scientists and engineers were recruited, trained, and rewarded on the basis of their ability to invent, while military officers were recruited, trained, and rewarded on the basis of their ability to wage war. Civilian scientists from outside the defense establishment had to be periodically introduced into the invention and innovation process to ensure the maximum rates of both.[2] Although Knorr and Morgenstern admitted that they could not prove their thesis, they pointed to a consensus that scientists were better at managing technological invention and innovation.

The argument that civilian scientists from outside the defense establishment have a critical and beneficial role goes back at least to World War II. During the years after the outbreak of war in Europe, American scientists led by Vannevar Bush organized and obtained presidential approval for the National Defense Research Committee (NDRC) and then the Office of Scientific Research and Development (OSRD).[3] Motivated by the failure of the National Research Council, an arm of the National Academy of Sciences, to provide any significant scientific support to the country during its brief involvement in World War I, Bush and other scientists pursued two objectives. The first was uncontroversial—to use the professional networks among scientists to mobilize them to work on military projects. Even military men who had wartime disputes with civilian scientists agreed that the NDRC, "being well loaded with university professors, was well able to mobilize the members of the union, something which . . . [the military] never could have done in a thousand years."[4]

[2]Ibid., pp. 13–16, 31–32.

[3]See, for example, Carrol Pursell, "Science Agencies in World War II, The OSRD and Its Challengers," in Nathan Reingold, ed., *The Sciences in the American Context: New Perspectives* (Washington D.C.: Smithsonian Institution Press, 1979), pp. 359–77.

[4]Harold G. Bowen, *Ships, Machinery, and Mossbacks: The Autobiography of a Naval Engineer* (Princeton: Princeton University Press, 1954), p. 178. Vice Admiral Bowen was the chief of the navy's Bureau of Engineering, which was in charge of the Naval Research Laboratory, and then in September 1939, after clashes with other naval officers, was demoted to head of the Naval Research Laboratory, where he clashed with OSRD over radar issues.

The other goal was far more profound in its implications. Science had revolutionized warfare as early as World War I, with the invention of poison gas and the airplane, but it was argued, military officers had been unable to properly utilize the contributions of science. Thus civilian scientists had to be given the dominant role not only in the invention of new ideas but also in the decisions about which ideas to develop and how the resulting weapons and systems ought to be used. As Bush wrote in a government memorandum in April 1942: "At the level of strategy there needs to be a control of trends in new weapons, a determination of emphasis, and an insistence on progress on specific matters at the expense of other things, if the situation is to be complete. The planning of military strategy needs to be carried on with a full grasp of the implications of new weapons, and also of the probable future trends of development."

Bush proposed, and President Roosevelt created, a Joint New Weapons and Equipment Committee of the Joint Chiefs of Staff. Bush was the chairman and the membership included representatives of the army and navy.[5]

The Office of Scientific Research and Development and the Joint New Weapons and Equipment Committee were not to confine themselves to mobilizing scientists and putting them at the disposal of the military. As the administrative history of the OSRD notes, it was the judgment of Bush and his associates that this approach had been tried and found wanting. "Previous efforts to bring civilian science into the program of weapons development were based on the theory that the Services would know what they needed and would ask scientists to aid in its development . . . [Now] the times called for a reversal of the situation, namely letting men who knew the latest advances in science become more familiar with the needs of the military so that they might tell the military what was possible in science."[6]

It would do no good, however, to introduce scientists into the process of military research and development unless they were allowed to retain their independence. Writing after the war ended, Bush ar-

[5]James Phinney Baxter 3rd, *Scientists against Time* (Cambridge: MIT Press, 1968), pp. 27–29. This book is the official history of the Office of Scientific Research and Development. The minutes of the sixty-four meetings of the Joint New Weapons and Equipment Committee are available in Record Group 218, Joint Chiefs of Staff, JCS Decimal No. 334, National Archives, Washington, D.C.
[6]Irwin Stewart, *Organizing Scientific Research for War: The Administrative History of the Office of Scientific Research and Development* (Boston: Little, Brown, 1948), p. 6.

gued that scientists working in military laboratories before the war had been treated as members of a

> lower caste of society by military officers, and that those labs, with a couple of notable exceptions (radar and sonar) had failed to produce any innovations. In the proper military-scientific partnership, scientists could not impose military plans on the military, since that was war planning which called on the professional expertise of the officer corps. By the same token, the military . . . [must accept] at the outset the principle that in synthesizing the judgments of diverse specialists into the integrated whole of a comprehensive plan, they will not override the professional judgment of others within the areas where those others have special competence.

By this he meant that scientists working on the development and selection of new weapons would accept no military interference. Eminent scientists, Bush asserted, "have no intention of being pushed around or placed in an inferior status, or of placing the judgments at which they arrive by the sweat of their brows before men of another profession for inexpert dissection or distortion."[7]

What was the basis for the claims made on behalf of an expanded role for the civilian scientific establishment? In the Anglo-Saxon world there was much evidence of bad relations between military officers and scientists in the period from 1914 to 1939. The Royal Navy had not bothered to establish any links with the scientific world to develop weapons with which to counter submarines before World War I began. In response to this failure, the committee on Imperial Defense established in June 1915 an independent Board of Invention and Research (BIR), composed of distinguished civilian scientists under the chairmanship of Lord Fisher (after his resignation as First Sea Lord after the failure of the Dardanelles expedition). Fisher gave the development of weapons to counter the German submarine and airship menace top priority. But scientists and military men clashed immediately. The Royal Navy refused to make available to the board any of its own submarines for the development and field testing of new devices. A Royal Navy officer who did obey the orders of the civilians of the board was in punishment confined to his quarters by the Royal Navy. The navy also blocked the production of a hydrophone developed by the board without any

[7]Vannevar Bush, *Modern Arms and Free Men* (New York: Simon and Schuster, 1949), pp. 19, 25, 253–54.

[225]

competitive testing and evaluation. The board itself was effectively forced out of existence in January 1918. Fisher's final judgment on his fellow naval officers and their response to civilian science was typically colorful: "We are doomed to exasperation and failure by not being able to overcome the pigheadedness of Departmental Idiots. . . . Never has the Admiralty Executive wholeheartedly supported the scientific and thoroughly practical proposals of the BIR research."[8]

In the United States, the German submarine threat prompted government leaders to reach out to the scientific world for aid in the development of countermeasures. After the sinking of the *Lusitania*, Secretary of the Navy Josephus Daniels sought out Thomas Edison and asked his help in bringing the navy together with the best academic and industrial research talent. The result was the Naval Consulting Board, which began operations in October 1915 with the object of setting up a laboratory. A deadlock emerged immediately, however, when Edison demanded that a civilian be placed in charge of the proposed lab and that the lab itself "work under civilian conditions away from naval and government conditions." The lab should be "purely civilian" and "not . . . have anything to do with the Navy except that if any naval officer has an idea he can go there and have it made." Relations grew worse when in January 1916 electrical batteries developed by Edison's own labs for use on U.S. submarines were involved in an explosion that killed four men. Edison's assistant wrote that after this accident the navy accused Edison and the assistant of "having sold something to the Navy Department that is a gold brick and being a pair of crooks not worthy to be trusted with the confidential relations that members of the Naval Consulting Board should and must bear to the Navy Department if any results are to be achieved by the Board." Edison withdrew from the project, and his final judgment on the scientific-military partnership resembled Fisher's. "I made about forty-five inventions during the war, all perfectly good ones, and they pigeon-holed every one of them. The Naval officer resents any interference by civilians. Those fellows are a close corporation."[9]

The fact that relations between the military and scientific communities had been bad did not necessarily mean, however, that putting

---

[8]Cited in Willem Hackmann, *Seek and Strike: Sonar, Anti-submarine Warfare, and the Royal Navy, 1914–1954* (London: HMSO, 1984), pp. 11, 16–37.

[9]Cited in David Kite Allison, *New Eye for the Navy: The Origin of Radar at the Naval Research Laboratory*, NRL Report 8466 (Washington, D.C.: Naval Research Laboratory, 29 September 1981), pp. 17–31.

the scientists in charge would result in superior research and development programs. We have seen the difficulty of accurately projecting the costs and military value of research and development and the difficulty of obtaining and analyzing intelligence about the enemy. What was it about the new role for scientists advocated by Bush that would improve the effectiveness and efficiency of technological innovation in the American military? Bush made a number of assertions that reflected the history of unfortunate relations between the two communities. "Really striking movements away from the crowd will not occur once in a thousand years in a tight bureaucracy." Yet that bureaucracy did, by his admission, produce radar and sonar in the twenty years between the wars. "There is something about the word ordnance that produces stodginess in its adherents." But might there not be something about the word professor that produced impracticality in its adherents? Why would scientists do better?

To be sure, civilian scientists knew science, but that did not necessarily mean that they understood military applications better than military men or scientists working in military laboratories. Bush himself is an interesting case in this regard. He emerged from World War II as one of the two or three most eminent men of science with responsibilities in the American military, and he wielded enormous influence. In a book he wrote in 1949 he made public many of his analyses and recommendations. Were intercontinental ballistic missiles possible and strategically rational? Bush stated clearly that they were not. "The question is particularly pertinent because some eminent military men, exhilarated perhaps by a short immersion in matters scientific, have publicly stated that they are. . . . We even have the exposition of missiles so fast that they leave the earth and proceed about it indefinitely as satellites . . . for some vaguely specified scientific purpose. All sort of prognostications have been pulled from the Pandora's box of science. . . . [T]hey have done much harm." Would the invention of nuclear weapons produce a revolution in warfare? No, for several reasons. First, atomic bombs were going to remain scarce and expensive "for a long time to come." Jet engines used too much fuel to be suitable for use on long range bombers, so that the defense, in the form of high speed jet-propelled fighters armed with guided missiles and guided by better radar networks, would have the edge over the few, slow, nuclear armed bombers. Bush wrote that if his analysis was correct, "we have less reason to be terrified by the thought of the A-bomb delivered by fleets of bombers. . . . There is a defense against the atomic bomb. It is the same sort of defense used against any other type of bomb." The era of mass bombing might be over. As for television-guided bombs and miniature atomic weapons fired by artillery or rocket, these were "Buck Rogers" fantasies.[10] Given what we now know of the developments of the next ten years, Bush cannot claim to have had in 1949 a clearly superior view of the possibility and utility of technological innovations.

What other arguments could Bush marshall in support of his proposed new role for independent, civilian scientist? After the war, Bush and others associated with the OSRD made a less polemical argument: military organizations were hierarchies that were "well adapted to the conduct of military operations and to the production and methodical improvement of standardized equipment. They are anything but favorable to the conditions under which scientific inquiry best thrives." Bush elaborated on this point in testimony before the Congress in 1945:

> Basically, research and procurement are incompatible. New developments are upsetting to procurement standards and procurement schedules. A procurement group is under the constant urge to regularize and standardize, particularly when funds are limited. . . . Research, however, is the exploration of the unknown. It is speculative, uncertain. It cannot be standardized. It succeeds, moreover, in virtually direct proportion to its freedom from performance controls, production pressures, and traditional approaches. . . . To be effective, new devices must be the responsibility of a group of enthusiasts whose attentions are undiluted by other and conflicting responsibilities.[11]

When organizations were made responsible for both research and procurement, research would suffer, and innovations would be slow in coming. This argument has some empirical support. There is considerable evidence of tension between officers responsible for military procurement and those officials, scientists or nonscientists, responsible for research. The chief of the Ordnance Department Technical Service, Colonel Gladeon Barnes, argued in 1940 that as production increased to handle arms exports to Great Britain and the effort to rearm the American military, a separate research divi-

---

[10]Bush, *Modern Arms and Free Men*, pp. 81–89, 93, 100, 101–2, 106.
[11]Testimony of Vannevar Bush before the Select Committee on Post-War Military Policy, 78th Congress, 2d session, 26 January 1945, cited in Baxter, *Scientists against Time*, p. 12. See also Baxter, p. 7.

sion in the department became more and more necessary. For his pains, Barnes was transferred out of the Technical Service and into production.[12]

Would civilian scientists acting outside the normal bureaucratic constraints be better able to avoid the harmful tension between research and procurement than their military predecessors? OSRD was set up to circumvent normal bureaucratic practice. Scientists were retained on contract, thus avoiding the constraints of civil service hiring, and were not commissioned into the army or navy. They were therefore responsible only to the scientists who headed the various divisions within OSRD, which was itself responsible to the NDRC, which operated under presidential charter, not under the War Department or Navy Department.[13] How much better did OSRD do?

In terms of the tension between procurement and research, it is clear that OSRD was no more free than any branch of the military from the pressures to sacrifice longer term research for the sake of short term procurement. There was, after all, a war going on, and the purpose of OSRD was to help win that war, not to lay the basis for postwar military-technological superiority. OSRD placed considerable emphasis on the production of weapons for field use. What is striking about the administrative and financial history of the OSRD is how much activity and resources went into procurement and support of equipment in the field, and not into basic or applied research that promised longer term payoffs. It is equally striking how much of the effort of the OSRD resulted from requests from the military, rather than from scientists "telling the military what was possible in science."

Wartime demands gradually pushed the OSRD into the business, not of research, but of production. The divisions of OSRD with the largest budgets, Division 3, in charge of solid rockets, and Divisions 14 and 15, responsible for radar and electronics countermeasures, were the primary sources of operational equipment in these areas for which there was no industrial base or in which relatively small numbers of complex components were needed. Together, these divisions took $204 million out of a total $457 million spent on contract research during the war. Bush himself was the catalyst for this transformation of OSRD from a research to a procurement agency;

[12]Ibid., p. 85.
[13]Stewart, *Organizing Scientific Research*, pp. 59–60, 80–83.

in November 1941 he, financier Elihu Root Jr, and industrialist F. S. Gordon formulated a plan to produce for immediate field use items then available only in the labs. This led to the creation of the Transition Office, to facilitate the production of items for limited service procurement and to dispatch scientists to the field to service new equipment.[14]

Scientists also needed to be in the field to measure the combat performance of new equipment and to refine and adjust the new technology and the operating practices developed for it. In the field they might also gain information about enemy activity that might be useful in subsequent weapons development. In October 1943 the Office of Field Services was created under OSRD. Originally called the Operations Analysis Division, it sent 464 scientists to field commands to help the navy shake down the first shipboard Combat Information Centers, to help develop antisubmarine warfare equipment and tactics, and to refine the electronics warfare capabilities of the Army Air Forces.[15] The activities of the Office of Field Services constitute one of the outstanding contributions of the OSRD, but it was an operation geared to help the military deal with its current operational procurement needs, not with pushing back the boundaries of science or revealing to the military the military possibilities of new technology. This is not to suggest that OSRD was in any way mistaken in its shift of emphasis. But it does demonstrate that even an organization staffed and headed by civilian scientists, protected by a presidential charter, and operating outside the bounds of the military bureaucracy could not refuse to take procurement priorities more seriously than those of research. Introducing civilian scientists into the picture did not remove or neutralize the pressures created by procurement needs.

Were there other reasons why civilian scientific establishments should perform notably better than those branches of the military charged with research and development? During World War II, a British physicist doing operations research argued that the training of scientists, engineers, and soldiers differed in ways that made scientists best able to interpret military data in order to assess the need for new technologies. "The soldier's training teaches him . . . [the habit] of forming correct snap decisions on inadequate data, and this habit acquires such importance that he still tends to use it on

---

[14]Ibid., pp. 66–67, 74–77, 85, 92–93, 232.

[15]Lincoln R. Thiesmeyer and John E. Burchard, *Combat Scientists*, volume in series Science in World War II (Boston: Little, Brown, 1947), pp. 27, 58, 108–9, 112, 113–16, 258–61.

occasions when it is wrong to do so." Thus, the soldier will observe one test of a new weapon on a training range or in the field and make up his mind if it is good or bad. Engineers were not better. They were trained to build systems, like bridges or airplanes,which should work all the time. They tended not to understand that a new technology that was effective in combat 30 percent of the time might represent a great success. Only physicists, and in particular nuclear physicists trained in probability theory and accustomed to "squeezing information out of rather unreliable material," had the proper background for assessing the need for and utility of new military technologies.[16]

More recently, academics have argued that independent scientists will better manage technology because military organizations resist the introduction of innovative technologies "that lie beyond their existing areas of technical expertise, and that either perform new missions or carry out existing ones in novel ways. They are far more prone to creating follow-on weapons that perform established missions in familiar sorts of ways." The U.S. Air Force, it is claimed, fought the introduction of the ICBM, and the U.S. Navy "bitterly fought the development of the first modern . . . cruise missiles."[17] The reasons, ostensibly, are straightforward. The Air Force resisted the ICBM because it was "committed to manned aircraft, and particularly to manned bombers, and refused to change. . . . Men who had always flown and relied upon bombers found it hard, indeed almost impossible, to sense the revolutionary implications of ballistic missiles."[18] The bureaucratic imperative to preserve existing missions and ways of operating tends to crush the impulse to make technological innovations.

Vannevar Bush made the same point when he wrote that self propelled artillery, rockets, recoilless guns, guided missiles, proximity fuses, and bazookas all were neglectedby the peacetime American military and had "waited for the pressure of war, appearing then largely outside the organized system of ordnance development, and

---

[16]2 June 1942 report by Dr. C. G. Darwin, correspondence of the Joint New Weapons and Equipment Committee, Joint Chiefs of Staff, Record Group 218, JCS Decimal No. 334, National Archives, Washington, D.C.

[17]Paul Noble Stockton, "Services and Civilians: Problems in the American Way of Developing Strategic Weaponry " (Ph.D. diss., Harvard University, 1986), pp. 5–6 and abstract.

[18]Edmund Beard, *Developing the ICBM: A Study in Bureaucratic Politics* (New York: Columbia University Press, 1976), p. 8. Beard notes that his argument is a simplification but does not seriously modify it in his book.

sometimes in spite of it."[19] Subsequent historians and social scientists have also pointed out how Bush took the development of radar during World War II away from the Naval Research Laboratory (NRL) and put it into its own, civilian, OSRD division. This was necessary, it is alleged, because the head of the NRL resisted the introduction of microwave radar. He was supposed to have favored the longer-wave radars pioneered by the NRL during the interwar period, in a classic case of the "not invented here" syndrome.[20]

This range of systems listed by Bush represents a cross section of new military technologies. The proximity fuse, microwave radar, and guided cruise and ballistic missiles, it can reasonably be argued, represent the new technologies that, setting aside the special case of nuclear weapons, most changed the character of warfare in this period. A review of the history of the development of these systems provides a way to determine how the American military reacted to innovative technologies, and to examine the role of civilians in helping the services to overcome the internal obstacles to technological innovation.

A look at some of the more minor technologies suggests that the picture is not as simple as the one painted by the scientific community. In the case of self-propelled artillery, a review board was created by the chief of staff of the U.S. Army, General Peyton Marsh, in December 1918 to review the experiences of World War I, and to determine the future artillery requirements of the army. It became known as the Caliber Board, or the Westervelt Board, after its chairman, Brigadier General William Westervelt. Its May 1919 report stated: "Mechanical transport is the prime mover of the future. . . . The first country utilizing the new capabilities opened up by mechanical traction and the caterpillar will have a great advantage in the next war." This gave the Ordnance Department a development agenda, and by 1941, it had developed all of the artillery pieces used in World War II, with the exceptions of a heavy antiaircraft gun and two field guns. The problem was that the combat arm that used the artillery, the Field Artillery, continued to argue that towed artillery was more maneuverable, less conspicuous, and more reli-

---

[19]Bush, *Modern Arms*, p. 25.

[20]Harvey M. Sapolsky, "Academic Science and the Military: The Years Since the Second World War," in Reingold, *The Sciences in the American Context*, pp. 382, 393; Daniel J. Kevles, "Scientists, the Military, and the Control of Postwar Defense Research: The Case of the Research Board for National Secruity 1944–1946," *Technology and Culture*, 16 (January 1975), 23.

able than motorized artillery.[21] There was a serious organizational obstacle to technological innovation, but resistance did not arise from army labs taking a "not invented here" position. It was the doctrinal preferences of the combat arms that created the difficulty. There is no obvious reason why civilian scientists would be more or less effective than military engineers in telling artillery officers in combat commands that their ideas about the tactical employment of artillery were incorrect.

In the case of recoilless rifles, the record indicates the Small Arms Divison of the Ordnance Department began work on these weapons in early 1943 on its own initiative after examples were captured from the Germans. A prototype weapon was ready in November of that year. The Joint New Weapons Committee, which was chaired by Bush, first took notice of the technology when British knowledge of this weapon was reported on 5 May 1943.[22] The Germans had clearly been ahead of the United States in this area, but the development branch of the U.S. Army and the civilian scientist heading new weapons programs responded at roughly the same time, with the Ordnance Department perhaps being slightly ahead of Bush. In the case of the antitank weapon, the bazooka, it is clear that the Ordnance Department was guilty of grave complacency in overlooking the basic technologies of shaped charges and solid rockets for field use, technologies that had been patented and demonstrated to the Army in 1911 and 1918 respectively. On the other hand, the Ordnance Department did produce a working model of a shoulder-launched rocket with a shaped charge for antitank use in November 1941, before the NDRC stepped in in April 1942 to assist in the further development of the launcher. Many military men argue that the bazooka should have been much more effective, but it should be noted that captured German officers expressed considerable reservations about their own weapon, the *Panzerschreck* or *Panzerfaust*, and the Soviet Union ordered the American bazooka as soon as it was demonstrated to its representatives.[23] Once again, the performance of the military engineers left much to be desired, but it is not clear

---

[21]Constance McLaughlin Green, Harry C. Thomson, and Peter C. Roots, *The Technical Services and the Ordnance Department: Planning Munitions for War* (Washington, D.C.: GPO, 1955), pp. 169–72, 314–17, 325–26.

[22]Ibid., p. 331; minutes of the thirty-first meeting, 5 May 1943, Establishment of Joint New Weapons Committee, Joint Chiefs of Staff, JCS Decimal No. 334, Record Group 218, National Archives, Washington, D.C.

[23]Green, *The Ordnance Department*, pp. 212–13, 329–30, 356–58.

that civilian scientists did play or could have played a major role in rectifying their errors.

The proximity fuse, which utilized a simple radar that could withstand the forces of artillery firing and could detonate the charge either when it came close to an airplane or when, after completing its trajectory, it came close to the ground (to kill and injure troops with shrapnel from above) was a major technological innovation never matched by the Germans. It played a major role in defending against the V-1 "buzz bomb" offensive against London, Antwerp, and Liège and in defeating German ground forces in the Battle of the Bulge. It was one of the major accomplishments of the OSRD and represented an accomplishment far beyond the capacity of the Ordnance Department. But while it took place outside the normal bureaucracy, it did not take place despite it. The scientist in charge of the program, Merle Tuve, pointed out that no one, in or out of the military, had done any work on proximity fuses until the Ordnance Department requested the National Academy of Sciences to look into the problem in August 1940. The decisive engineering breakthrough was the development of small, rugged vacuum tubes, apparently an unintended benefit of research in the hearing-aid industry. The crucial circuit design breakthrough was made by the British.[24] All due credit should go to the scientists involved, but it was a clear case of the military requesting new technology that it could not invent itself, not of the scientific community inventing something the military had not imagined.

The case of microwave radar is also instructive.[25] Despite the quarrels with Edison, the navy did set up the Naval Research Laboratory (NRL), with a navy officer in command, in 1919. Edison predicted: "If Navy officers are to control it the results will be zero. This is my experience due to association with them for two years and noting the effects of the system of education at Annapolis." NRL researchers discovered in 1922, as did many scientists around the world at this time, that ships and aircraft would interfere with radio waves and that this interference could be used to detect them at long distances. Additional research soon led to the discovery that both the ionosphere and objects at or near the surface could also reflect radio waves, and the distances of the ionosphere and of objects from the transmitter could be determined through radio pulses

[24]Ralph B. Baldwin, *The Deadly Fuze* (London: Jane's Publishing, 1980), pp. 57, 62, 65.
[25]The material in the following paragraphs draws heavily from Allison, *New Eye for the Navy*.

(as opposed to continuous transmissions). By 1930, NRL researchers were using radio waves to detect aircraft and determine their velocity. During this period, industrial scientists were preoccupied with satisfying the commercial demand for radios in the first consumer electronics boom in the U.S., so NRL scientists and engineers had the field of radar to themselves until at least 1933.

All of the initial work was done with high frequency radio waves, not microwave radiation. These efforts lead to the successful development of radars for shipboard use. An NRL designed radar was chosen after a competition with a radar designed by RCA. The use of the NRL radar on U.S. battleships in exercises in 1937 provided such dramatic benefits in defending those ships against nighttime attack that, in the words of one of the NRL researchers, "from then on, they [the operational Navy officers] were sold on the stuff and they would give us anything we wanted." In 1939, the chief of Naval Operations ordered radar for installation on cruisers and battleships of the fleet.[26]

In terms of the debate concerning the origins of military technological innovation, the key innovations were made by scientists working under navy command. Delays due to budget constraints did occur while basic research was underway, but the demonstrated operational benefits of the new technologies were such that the innovations received enthusiastic support, not opposition. Demonstrable results were the critical factor, not the origin of the technology. Civilian scientists working for the navy were just as interested as other scientists in pursuing any scientifically interesting research, but persuading the navy to provide major funding before results could be shown was a problem. Uncertainties about the value and costs of technological research are a universal problem, not confined to military or civilian researchers and managers.

Did the NRL overlook the benefits of microwave radar because it had pioneered radar with lower frequencies? The answer seems to be no. The theoretical benefits of microwave radar (superior resolution and ability to distinguish among types of targets, smaller antennae, which made airborne radar much more feasible) were well understood by the NRL and Army Signal Corps, and the NRL successfully appealed for more money for microwave research from Congress in 1935. But vacuum tubes that could generate microwave signals of sufficient strength for military use could not be produced. Microwave radar was thus dropped by the NRL as a

[26]Ibid., pp. 109–11.

[235]

development project in 1937, pending development of the necessary tubes. Harold Bowen, then director of the NRL, did not forget microwave radar, however, reminding the secretary of the navy in a letter written in December 1939 of the potential of microwave radar for use on Navy bombers.[27]

The British had also recognized the potential of microwaves and the need to develop new technologies for producing them in sufficient strength. In a genuine burst of inspiration, a scientist at Birmingham University, J. T. Randall, developed a solution that completely bypassed conventional theories of radio wave propagation. His prototype device, the multicavity magnetron, was demonstrated in 1939 and produced orders of magnitude more power than earlier magnetrons in either the United States or Great Britain. The Tizard scientific mission from Great Britain to the United States in August–September 1940 brought the multicavity magnetron with it as part of a broader exchange of military technologies, and it was eagerly embraced by the NRL, which wrote a memo to the secretary of the navy explaining the revolutionary implications of the British device.[28] It was not U.S. civilian scientists who invented the microwave radar and forced it on the NRL. Nor was this a case in which an independent civilian scientist outperformed the U.S. defense establishment. Randall began work on his device only after he had been recruited in the autumn of 1939 into a consortium of scientists and electrical engineering firms organized and coordinated by the Director of Scientific Research of the British Admiralty.[29]

Harold Bowen, the director of the NRL during this period, was an abrasive naval officer who, by his own admission, relished his next job of breaking strikes in defense plants during the war.[30] He was clearly not the man to bring scientists and military men together, and Bush was certainly correct in not allowing him to manage all radar programs during the war, as he sought to do. But the NRL had been able to develop radar, had sought out more advanced radar technologies and had been willing to accept them even when they came from another, foreign, lab. The breakthrough was made by a civilian scientist in Britain. In that country, as in the United States, technological innovation benefitted from wartime recruitment of talented individuals not otherwise interested in the military applica-

[27]Bowen, *Ships, Machinery and Mossbacks*, p. 147.
[28]Allison, *New Eye for the Navy*, pp. 151–152.
[29]J. G. Crowther and R. Whiddington, *Science at War* (New York: Philosophical Library, 1948), p. 45.
[30]Bowen, *Ships, Machinery, and Mossbacks*, pp. 137–204.

tions of science. This scientist was working in a program initiated by a military service to solve a problem defined by that service; he was not independently pointing the way to a reluctant military.

SCIENTISTS, MILITARY MEN, AND GUIDED MISSILES

A major new technology for unmanned weapons, one that would replace officers as combat commanders, would appear to be one of the most difficult for military organizations to impose on themselves, and thus one in which civilians might play a significant role. Analyses of the introduction of ballistic missiles into the U.S. Air Force and Navy have emphasized "the infusion of new non–Air Force personnel who could recognize" the changes in technology and the Soviet threat.[31] In particular, the role of the Assistant Secretary for Research and Development of the Air Force, Trevor Gardner,[32] and the civilian scientific teams he mobilized at Rand and in the von Neumann committee has been singled out. Senator Henry Jackson emphasized the navy's reluctance to divert money from traditional missions to the new Polaris submarine-based ballistic missile. "The result," Jackson said, "was that Polaris was not pushed hard until Sputnik came along."[33] We have seen, however, that skepticism about strategic missiles was not confined to the military; that Vannevar Bush also opposed ballistic missiles after World War II. A detailed examination reveals a more complicated picture than that of civilian advocacy and military resistance.

The U.S. Army and Navy became interested in unmanned guided missiles for bombardment in 1916, when the civilian inventors Elmer Sperry and Peter Hewitt developed an automatic control mechanism that could be installed on an aircraft to convert it into an unmanned flying bomb. The navy provided $200,000 in funding for prototypes in 1916, and observers at the first trials noted that its primary value would be in attacking large urban areas, where its lack of accuracy would not matter and where it might have the greatest psychological impact on the enemy. The initial goal was to mass produce these primitive cruise missiles at a cost of $2,500 a copy. The army funded

---

[31]Beard, *Developing the ICBM*, p. 4.

[32]See, for example, Michael H. Armacost, *The Politics of Weapons Innovation: The Thor-Jupiter Controversy* (New York: Columbia University Press, 1969), p. 57.

[33]Ibid., p. 66, citing U.S. Congress, Senate, Committee on Government Operations, *Hearings, Organizing for National Security* I, 87th Congress, 1st session, 1961, pp. 1084–85.

a separate project in which a young army officer, Hap Arnold, worked with Charles Kettering to develop the "Kettering Bug," an unmanned drone designed to be inexpensively produced in numbers up to 100,000 units. Both the army and navy programs languished after World War I due to a lack of funds and problems in designing an aerodynamically stable unmanned vehicle. Drones did not become cheap substitutes for manned aircraft. Because of the limitations of radio controls, they needed to be accompanied by a controlling airplane, and remained as expensive as a manned military aircraft. Because they had to be observed by the controller, they could be flown only in good weather during the day.[34] However, work on the technology did continue.

The need for target drones, as opposed to assault craft, persisted, and here cost was not so much a factor, since relatively few target planes were needed. The navy continued work with radio-controlled drones for this purpose, developing guidance mechanisms virtually identical to those used by the Germans for their guided bombs used in World War II. The navy noticed in 1940 that its antiaircraft gunners were, on repeated occasions, unable to shoot down the drones during target practice. The relative invulnerability of these craft spurred interest in their use for combat. The navy began a series of development programs for drones for antiship missions, and the Naval Research Laboratory developed television and radar sensors that could be carried on the drones to provide target data that would be used by a controller. A television guided missile successfully hit a towed target in April 1942, and radar-guided missiles proceeded into development. In 1943, the navy approved the purchase of two-thousand assault drones, training programs for operators were instituted, and operational units formed. However, manned aircraft for assault missions were also being developed, and they were having considerable success in combat in the Pacific. The assault drone program would have utilized 10,000 men, including 1,300 aviators in the control planes, and would have diverted considerable production capacity from proven weapons systems. It was not, therefore, cost free, nor was it clearly superior to the manned alternatives. The program was essentially killed in September 1944 by the head of the navy, Admiral King. "War weary" bombers that could no longer be used for manned missions did not impose the

---

[34]Kenneth P. Werrell, *The Evolution of the Cruise Missile* (Maxwell Air Force Base, Ala.: Air University Press, 1985), pp. 7–16, 19, 23.

same costs, and first-generation guided missiles based on these converted bombers were used in combat against Heligoland, Bougainville, and Rabaul, with some success.[35]

The U.S. Army Air Forces had also been sponsoring work by its Experimental Engineering Section at Wright Field during the 1930s. When queried by the Joint New Weapons and Equipment Committee in May 1942 about the status of "controlled missile" programs,[36] Wright Field was able to report that a number of important guidance components, including triple-axis gyrostabilizers, servomotor controls, and radio control links, had been developed and tested. Drones for target purposes were on hand, and propeller-driven cruise missiles with radio controls carrying five-hundred-pound payloads a range of seven-hundred miles were on order from General Motors. Unpowered guided glide bombers, television-guided and other dirigible bombs were also in development, all utilizing the control mechanisms developed at Wright Field during the 1930s.[37] When asked in December 1942 what the Army Air Forces research and production priorities were for "controlled" missiles, Major General George Stratemeyer replied that the first was to produce an aerodynamically stable design for unpowered radio-controlled glide bombs, and to develop television and radar "systems of intelligence" (i.e., guidance systems) for them. The second priority was to develop airframes for unmanned powered flying bombs with payloads of two-hundred pounds and ranges of one-thousand miles. The head of the OSRD division in charge of guided missiles, H. B. Richmonds, informed the Joint New Weapons Committee that there was no existing analysis in the scientific community that contradicted the priorities set by the Army Air Forces.[38]

---

[35]Louis A. Gebhard, *Evolution of Naval Radio-Electronics and Contributions of the Naval Research Laboratory* (Washington, D.C.: Naval Research Laboratory, 1979), pp. 227–29, 232; Werrell, *Evolution of the Cruise Missile*, pp. 25–26.

[36]Minutes of the first meeting, 12 May 1942, Joint New Weapons and Equipment Committee, Establishment of Joint New Weapons and Equipment Committee (hereinafter JNW), Joint Chiefs of Staff, Record Group 218, JCS Decimal No. 34, National Archives, Washington, D.C.

[37]Report by Colonel F. O. Carroll, Air Corps, Chief, Experimental Engineering Section, Wright Field to the Commanding General, Army Air Forces Material Command, 19 May 1942, copy forwarded to the JNW, 25 May 1942, Controlled Missiles Project (hereinafter CM), Joint Chiefs of Staff, Record Group 218, JCS Decimal No. 471, National Archives, Washington, D.C.

[38]Stratemeyer to Army Chief of Staff for Logistics G-4, "Re: Controlled Missiles," 4 December 1942; H. B. Richmond to JNW, 9 December 1942, CM, JCS, Record Group 218, JCS Decimal No. 471, National Archives, Washington D.C.

When the Joint New Weapons Committee revisited the guided missile programs of the services in 1943, they found short-range guided glide bombs, longer-range guided glide bombs, television- and radar-guided bombs, photoelectric homing devices and power-driven bombs all in development. All would be launched from manned bombers but were unmanned for all of their flight.[39]

The next year, in response to the German jet-powered flying bomb, the V-1, the Army Air Forces instituted a program to build a copy, the JB-2. In July 1944, the Air Forces ordered 1,000 JB-2s, and proposed the production of 1,000 a month by April 1945 and 5,000 a month by September 1945. In January 1945, the chief of staff of the Air Forces actually ordered 75,000 JB-2s for use in Europe, and ordered the production of 100 a day by September, and 500 a day in a year's time. These plans proved rash. Subsequent analysis showed that if these plans were executed, artillery production would have to be cut 25 percent, bomb production by 17 percent. The shipping space necessary to carry the JB-2s to Great Britain was necessary for other critical war materials. At the same time, the eighth and fifteenth Air Forces were reaching the climax of their bombing operations. By April 1945, they had left very little in Germany to bomb. Plans for the JB-2 were cut back, and only 1,400 were actually procured.

The Army Air Forces had big postwar plans for cruise missiles. The deputy chief of staff of the Army Air Forces wrote to Vannevar Bush in February 1945: "We believe the JB-2 to be representative of a new family of very long range weapons whose capabilities will profoundly affect future warfare and especially aerial warfare. We want now to explore the possibility of very long range missiles to the utmost extent which will not involve a serious diversion of effort from the essential business of prosecuting this war."[40] Nor was the navy idle. A review in early 1944 found it working on assault drones with television and radio remote control, radar homing devices for drones, glide bombs with television controls (Glomb), the Gorgon jet powered, radio controlled antiship and antiaircraft missile, which was to be tested in April 1944, and infrared sensors for guided missiles.[41]

---

[39]24 May 1943 memorandum of the secretary of the JNW to JNW members summarizing controlled missile programs, CM, JCS, Record Group 218, JCS Decimal No. 471, National Archives, Washington D.C.

[40]Werrell, *Evolution of the Cruise Missile*, pp. 63–65, 79.

[41]4 February 1944 draft report of the JNW to the JCS on army and navy guided missile programs, "Enclosure B, Navy Programs," CM, JCS, Record Group 218, JCS Decimal No. 471, National Archives, Washington, D.C.

Much of this work involved missiles that would be launched from manned aircraft. Were the services biased in favor of systems that complemented rather than replaced manned aircraft? If so, the research and development priorities of the civilian NDRC Division 5, with which the services worked during World War II on guided missiles, showed the same bias. The classified report of Division 5 published in 1946 focused exclusively on four bombs, all launched from manned aircraft, the Razon and Tarzon guided bombs, which were visually directed to the target after release from heavy bombers or fighter-bombers, the Felix heat-seeking bomb, and the Roc radar-homing bomb.[42]

In summary, far from being opposed to unmanned guided missiles, the Army Air Forces and the navy sponsored research during the 1920s and 1930s that led to the guided weapons used in World War II. When Vannevar Bush began to push guided missiles in 1942 from his position on the Joint New Weapons Committee, he found numerous guided weapons programs already in existence. The services pushed them aggressively during the war, with important technical support from civilian scientists, but without significant civilian direction. Bush summarized the relationship of the service programs to civilian scientific direction by complaining in 1944 that there were too many service-initiated guided missile projects that had not been effectively controlled by the Joint New Weapons Committee. He found thirty-eight separate programs, which he thought to be "over-elaborate," and he sought to institute a "well balanced program aimed at specific objectives." He was frustrated in this effort, since there were many groups having projects for which they had great enthusiasm and that each group was able to persuade its service to support it.[43] Bush found this multiplicity of programs "absurd," and it may well have been so. This plethora of programs, however, simply cannot be reconciled with the conventional picture of services dominated by the pilots of manned aircraft who would have strangled unmanned missiles programs were it not for the intervention of civilian scientists.

When World War II ended, the Air Force continued to move toward guided missiles. Hap Arnold led the move *away* from manned aircraft when in 1945 he addressed the von Karman committee:

[42]Hugh H. Spencer, Chief, *Summary Technical Report of Division 5, NDRC, Volume I, Guided Missiles and Techniques* (Washington, D.C.: Office of Scientific Research and Development, 1946), pp. 1–4.

[43]Vannevar Bush to JNW, 29 August 1944, Guided Missiles Correspondence, JCS, Record Group 218, JCS Decimal No. 471.6, National Archives.

"For twenty years the Air Force was built around pilots, pilots, and more pilots. The next twenty years is going to be built around scientists."[44] Arnold's assumption was that the American people would not tolerate a large standing military or wars in which manpower losses were heavy. Machines would thus increasingly have to be substituted for military manpower. Arnold's judgment was seconded by General Curtis LeMay in a memorandum to Arnold's successor, General Carl Spaatz, in September 1946. LeMay, the Air Force's assistant chief of staff for research and development, also emphasized the need to substitute missiles for manned bombers because of anticipated casualty rates: "The long range future of the AAF lies in the field of guided missiles. Atomic propulsion may not be usable in manned aircraft in the near future, nor can accurate placement of atomic warheads be done without sacrifice of crews. In acceleration, temperature, endurance, multiplicity of functions, courage, and many other pilot requirements, we are reaching human limits. Machines have greater endurance, will stand more ambient conditions, will perform more functions accurately, will dive into targets without hesitation. The AAF must go to guided missiles for the initial heavy casualty phases of future wars."[45] LeMay's memo does not mention the fact that in the 1940s, nuclear weapons were scarce, numbering in the tens. For the foreseeable future their use in strategic bombing campaigns would have to be supplemented by numerous manned bombers carrying nonnuclear weapons.[46] Guided missiles carrying nuclear warheads, if feasible, could replace manned bombers in attacking the most difficult targets in the initial phase of war. The scarcity of nuclear weapons made necessary a commitment to both manned and unmanned strategic systems.

In the longer term, guided missiles would play a larger, if still uncertain role. The commitment to guided missiles in the Air Force was backed up by money and programs. Out of the thirty-eight guided-missile programs sponsored by all U.S. services at the end of World War II, nineteen were Army Air Force programs. This number grew to twenty-one in January 1946, and to forty-seven by mid-1946. In 1946 the Air Force set aside 2.6 percent of its R&D budget for guided missiles, and planned to see that share grow to 34.4 percent by fiscal year 1947.[47] This trend did not materialize as postwar

---

[44]Quotation from unpublished memoirs of a member of the von Karman team, cited in Michael Gorn, *Harnessing the Genie* (Washington, D.C.: GPO, 1988) pp. 17–18.

[45]Beard, *Developing the ICBM*, p. 39.

[46]See previous chapter, p. 185.

[47]Werrell, *Evolution of the Cruise Missile*, p. 81; Beard, *Developing the ICBM*, p. 52.

defense budgets were slashed, and the Air Force was reduced to three missile programs by 1949, all cruise missiles.[48] But manned bombers were severely cut back as well. General Hoyt Vandenburg said in 1949 that after World War II, "everything went downhill so fast that the first thing we had to pay attention to [in 1948 during the Berlin crisis] was to get a sort of fire-bucket brigade ready in case something should break." That bucket brigade was the handful of B-29s modified to carry nuclear bombs dispatched to Great Britain in 1948. Now, Vandenberg continued, "we have gotten our people together to the point where we feel we have a force in being; therefore our thought naturally turns to 'where do we go from here?' "[49]

Managing Uncertainty

If civilian scientists did not have a clear and inherent advantage in the management of military technological innovation, what alternative strategies were there? One that had some appeal in the days when the United States had a relative abundance of resources available for military research and development was simply to build anything that looked promising, and then try it out. Herbert York reports, for example, how Secretary of Defense McElroy expressed his support for the development of the B-70 bomber in 1959: " 'I'm Secretary of Defense of the United States, the richest nation in the world and the leader of the West, and if we don't do it, no one will, so I'm for it.' "[50] While feasible if a nation is rich enough, the strategy has less appeal when hard choices have to be made.

The argument of this chapter and the one preceding is that the fundamental problem of managing military research and development is that uncertainties about the enemy and about the costs and benefits of new technologies make it impossible to identify the single best route to innovation. This logic suggests that it might be reasonable to give up the search for an optimum strategy and concentrate instead on ways of living with the uncertainties. If the future is uncertain, then it pays to be flexible. A strategy for military technological innovation that seeks as much flexibility as it can buy might be better than one of trying to buy the one weapon that

[48]Werrell, *Evolution of the Cruise Missile*, p. 81.
[49]Thomas A. Sturm, *The USAF Scientific Advisory Board* (Washington, D.C.: GPO, 1967), p. 33.
[50]Herbert York, *Making Weapons, Talking Peace* (New York: Basic Books, 1987), p. 180.

would perform the best if it could be built to specifications at the expected cost and if it eventually turned out to be the weapon which was actually needed.

The search for flexibility can easily turn into a search for weapons that will be useful in every possible contingency. Uncertainty can, indeed, be managed in this fashion. Economists have referred to this as Type I flexibility, and give as an example a shoe factory that can be used to produce every kind of shoe and is thus able to respond to the certainties of customer taste. The American military has purchased multipurpose weapons precisely because it is not certain about the conditions in which it will have to fight, and it has been willing to pay a price higher than the price for weapons optimized for one mission only. For example, aircraft carriers are more expensive than simple ships carrying cruise missiles, and manned bombers are more expensive than ICBMs. But aircraft carriers can be adapted to different conditions by being equipped with different kinds of airplanes, and manned bombers have proved to be useful in conflicts in which ICBMs have no utility at all. Alternatively, Type I multipurpose flexibility could be pursued by purchasing different weapons for every conceivable contingency and then creating different units structured around those contingencies. The costs of implementing this type of flexibility are clearly enormous.

In conditions of great uncertainty, when political and technological conditions are in flux, the pursuit of Type I flexibility may be impossible or prohibitively expensive. Under such conditions, Burton Klein has suggested the pursuit of Type II flexibility which "attempts to reduce the uncertainties confronting the decision maker by buying information on competing development alternatives. It is premised on the assumption that some of our resources can be used to reduce these uncertainties before military forces are actually procured and put on the line, that the greater knowledge attained by comparing development alternatives will contribute directly to widening the range of alternatives available and to reducing the number of uncertainties."[51]

Concretely, Type II flexibility manages uncertainty by buying information and then deferring large-scale production decisions. It buys information about the relative costs and performance of new technologies by investing money into the development of many

[51]Burton H. Klein, "Policy Issues in the Conduct of Military Development Programs," in Richard Thybout, ed., *Economics of Research and Development* (Columbus: Ohio State University Press, 1965), p. 324.

different weapons up to the point where their costs and military performance can be accurately assessed. This has usually meant bringing a weapon system to the prototype stage, where it can be tested and its performance accurately gauged. A few production weapons might be produced so that military units can experiment with them in field exercises and explore which doctrines make best use of them. But large-scale production is deferred, because political events or new technologies may come along that make the weapon more or less useful than was originally believed. Political circumstances, including the military technology of the enemy, may change and make the weapon more or less necessary. Large-scale procurement is deferred, in short to allow uncertainties to work themselves out. When long-term uncertainties become short-term requirements, decision makers can choose from an array of prototypes the system best suited for the needs of the day. A necessary component of this strategy, therefore, is a capacity for mobilizing production from prototypes.

A good example of this strategy in practice is provided by the long-range ballistic missile of the United States in the 1940s and 1950s. Postwar guided-missile program policy was dominated by major uncertainties. Not only were technologies in a period of rapid change, but U.S. alliances between 1945 and 1950 were rapidly changing from friendship with the Soviet Union, coupled with the general expectation of withdrawal of American soldiers from Europe, to hostility with the Soviet Union and a framework of peacetime military alliances unprecedented in U.S. history. Defense budgets were in flux. The question of how nuclear weapons would be used by the United States was unanswered. Congress had passed the MacMahon Act, which put control and development of nuclear weapons in the hands of the president and the Atomic Energy Commission. The chief of naval operations pointed out in 1951 that no decisions had been made by the civilian custodians as to whether nuclear weapons would be allocated for use on guided missiles. The Joint Chiefs had assumed that half of the available nuclear weapons would be so used, but the chief of naval operations believed that this was unrealistic.[52]

The most sensible plan for guided missile development was to hedge. In hindsight, an immediate decision in 1948 or 1950 in favor of the weapon system that did emerge as dominant, the

[52]11 December 1951 memorandum, Guided Missiles Committee, JCS, Record Group 218, JCS Decimal No. 334, National Archives, Washington, D.C.

ICBM, might appear reasonable. At the time, however, it was not at all clear which technologies for long-range strategic bombardment would be most useful, how or when they would be employed, or how they would best fit into existing service missions. In response to these uncertainties, the Guided Missiles Committee of the Joint Chiefs of Staff formulated a draft National Program for Guided Missiles. The committee argued that while certain very broad requirements could be imagined, such capabilities for the strategic bombardment of cities, for attacking small, hard military targets, for air defense, and for antisubmarine warfare, the unknowns were so large that it made sense to explore many different technologies but not spend money on procurement. Research money should go more for basic research and less for programs leading to specific weapons. Interservice battles about who would control what should be deferred until it is known what kind of wars the military would be called upon to fight, and what technologies would actually be developed.[53] This program was approved by the JCS on 22 March 1946 as JCS 1620, and a memorandum spelled out its implications:

1. Emphasis will be placed on further basic information in both fundamental and applied science.

2. Practical development is by far the most expensive part of the program. Consequently, practical development will not be rushed ahead of sound knowledge.

3. The desirability of competitive efforts on especially difficult problems will be recognized, subject to integrated overall consideration. . . Some duplication is valuable. . . . Military characteristics will be as flexible as possible.[54]

The Technical Evaluation Group of the JCS reviewed this basic policy in 1949. It did find some duplications of effort that it found unjustified, particularly in engine development, but it could find no wasted funds that could be better used elsewhere. Given what it could determine about technical opportunities, no field of potential development was being neglected. The group reaffirmed the wisdom of a broad research program that was not focused on weapons production. "The possession of scientific knowledge and engineer-

---

[53]21 November 1945, "A National Program for Guided Missiles," draft, Guided Missiles Committee, JCS, Record Group 218, JCS, Decimal No. 334, National Archives, Washington, D.C.
[54]22 March 1946, same file series as above.

ing techniques [will] prove more valuable in meeting the exigencies of a long war than a large stockpile of obsolescent missiles." While there was duplication of effort, "it is not clear that such a procedure increases the cost as much as is stated by superficial observers."[55]

This research and development plan was complemented by a mobilization plan for the large-scale production of missiles that were being developed in prototype. The Korean War created international political conditions in which large-scale war was plausible, and so plans were developed by the Joint Chiefs of Staff in October 1950 for the production of, for example, 800 medium-range ballistic missiles a month by June 1952 and 1,200 strategic surface-to-air missiles a month by September 1953.[56]

This 1945 policy of research instead of procurement, and its 1949 endorsement, also provided a hedge against an unexpected technological development—low-weight, high-yield hydrogen bombs, which made inaccurate ICBMs much more useful than could have been anticipated. All of the major participants in the development of the ICBM—from Bruno Augenstein, who headed the Rand team that prepared the report advocating an accelerated ICBM program in 1954, to the von Neumann committee, which was briefed by Augenstein in connection with its own similar report, to Air Force officer Bernie Schriever, who headed up the accelerated ICBM effort, to Air Force Secretary James Douglas—agree on one point: the critical event in the development of ICBMs was the development of the hydrogen bomb. Initial tests and calculations in late 1953 suggested that the ratio of warhead weight to explosive yield would plummet. This meant that warheads would be possible very soon that would be small enough and powerful enough that the payload and accuracy requirements of the Atlas ICBM, under development by the Convair Corporation as Air Force project MX-774 since 1946, could be relaxed from 3,000 to 10,000 pounds down to 1,500 pounds and from one-quarter of a mile to "two and probably three nautical miles."[57] The initial calculations of predicted hydrogen bomb perfor-

[55]20 May 1949 report of Technical Evaluation Group of the Research and Development Board, Guided Missiles Committee, same file series as above.

[56]Report of the Guided Missiles Interdepartmental Operational Requirements Group to the JCS, 6 October 1950, Guided Missiles Committee, JCS, Record Group 218, JCS Decimal No. 334, National Archives, Washington, D.C.

[57]"Recommendations of the Strategic Missile Evaluation Committee" (von Neumann committee), RW008-4, 10 February 1954, available from the Air Force Historical Division, Bolling Air Force Base, Va., p. 7; Bruno Augenstein, "A Revised Development Program for Ballistic Missiles," Special Memorandum No. 21 (Santa Monica, Calif.: Rand Corporation, 1954); Beard, *Developing the ICBM*, pp. 141–43.

mance were confirmed by the CASTLE test series carried out in 1954, and the recommendations of the von Neumann committee to create a new Air Force organization, the Western Development Division, to work with the Convair Corporation to build the Atlas ICBM were implemented in the same year. The easing of the technical requirements made the missile feasible in the mid-1950s instead of the early 1960s, as had been projected. Once the new design parameters had been laid down, the ICBM became an obvious choice, since its speed made it essentially invulnerable to the defenses of the time, unlike the manned bomber and the subsonic cruise missile.

It was not a Soviet threat, or a civilian scientific intervention in the context of fixed technological possibilities that pushed the innovation of the ICBM, but a new and unforeseen technological innovation created by civilian physicists. Because of the adoption of a strategy of buying hedges against uncertainty, the Atlas project had by 1953 solved all of the other major problems involved in ICBM development. More powerful engines, gimballed nozzles for directing thrust, and ultra lightweight designs for the fuel and oxidizer tanks had all been successfully developed before 1954.[58] The fact that a new administrative structure was created in 1954 to accelerate the Atlas program gave rise to the impression that Convair had done a mediocre job of development, but the retrospective judgment of Bruno Augenstein was that the people working on the Atlas at Convair "were . . . I think a very clever technical group . . . not . . . given the kudos that they deserved." Augenstein also stated that the head of Rand in 1954, Frank Colbaum, who was also involved in the ICBM review project, "felt that . . . a lot of industry people were in a sense being slighted by the notion that they had to have . . . this technical direction group set up under Trevor Gardner."[59]

In the navy, programs to produce the Regulus cruise missile that would be launched from submarines were underway from the 1940s on. It might be expected that officers involved in the cruise missile program would oppose submarine launched ballistic missiles, which would put the cruise missile out of business. But the earliest advocate of submarine launched ballistic missiles was Commander Robert Freitag, head of the Surface Launch Missile Branch of the Navy's Bureau of Aeronautics, who was in charge of the cruise missile program. In 1954, Freitag first suggested ballistic missiles for

[58]Beard, *Developing the ICBM*, pp. 63–64.
[59]Interview with Bruno Augenstein conducted by James Digby and Joan Goldhammer, 22 May 1986, Rand Corporation, Santa Monica, p. 16.

submarines. In 1955, he and his boss, the head of the Bureau of Aeronautics, Rear Admiral James Russell, began taking money allocated for the cruise missile to fund research on submarine launched ballistic missiles. Freitag received indirect support in 1955 from the commission of civilian scientists chaired by James Killian, but this was not decisive. There was opposition to the idea from within the navy, but the main battle had to be fought with the civilian deputy secretary of defense, Reuben Robertson, who resisted the creation of another ballistic missile program at a time when he was trying to centralize all such programs in the Air Force.

Two events were decisive in reversing the situation and creating an active Fleet Ballistic Missile program. First, Admiral Arleigh Burke became chief of naval operations in August 1955 and gave the idea of Fleet Ballistic Missiles his total support. Second, in 1956, discussions with Edward Teller during Project Nobska at the Woods Hole Summer Study made the navy aware of the implications of the new warhead technology; the Navy missile could be much smaller and lighter, and so a solid fuel rocket could be used. Until that time, it was believed that only a liquid fuel rocket could generate the necessary thrust, and the perils of storing liquid oxygen or nitric acid onboard a submarine had created serious technical problems. With a lighter warhead, the less powerful but much more easily handled solid rocket became a real option.[60] As in the case of the Air Force, an early strategy of buying technological hedges paid off when scientists developed unexpected new technologies. Scientists were effective not in overcoming organizational resistance or in setting policy, but in doing science.

CONCLUSIONS

The management of military technological innovation in the modern U.S. military appears to have been dominated by the problem of coping with uncertainties about the enemy and the costs and benefits of new technologies. The evidence from the periods before and after World War II does not show that the civilian scientific community had any inherent advantages over the military in the management of this uncertainty. Scientists made numerous, invaluable

[60]Harvey M. Sapolsky, *The Polaris System Development: Bureaucratic and Programmatic Success in Government* (Cambridge: Harvard University Press, 1972), pp. 7, 18–21, 29–32.

contributions to military technology. Military organizations made numerous errors in choosing which technological avenues to pursue, and military men clashed with scientists. But there is no evidence that civilian scientists had to intervene so that the military would make use of available technologies that were being neglected because of bureaucratic pathologies. Nor does the evidence suggest that civilian scientists would have necessarily done a better job of selecting which technologies to develop.

The evidence suggests that in the period 1945–1955 technological innovations in the U.S. military were largely the product of long-term development programs conducted within or at the direction of the military. Technological breakthroughs, such as microwave radar and the hydrogen bomb, did change parameters of innovations, but those breakthroughs were themselves products of programs designed explicitly to support the military, not of independent scientific investigations.

A successful strategy for managing uncertainty was developed and implemented by the U.S. Air Force and Navy in the area of guided missiles. Technological uncertainties were reduced by developing many different technologies to the point of procurement, but then deferring large-scale production while other uncertainties resolved themselves.

The overall picture of American military research and development in the period from 1930 to 1955 is one of technological innovation largely unaffected by the activities of potential enemies, a rather self-contained process in which actions and actors within the military establishment were the main determinants of innovation. In this sense the process resembles the peacetime and wartime organizational innovations discussed in earlier chapters, which were only loosely related to enemy capabilities and intentions. If this picture is correct, military innovation as a whole is much less bound up with foreign military behavior or civilian intervention than is ordinarily thought.

9

Conclusion:

Lessons Learned

This book began with a set of questions concerning the condition in which military innovation can take place, the relative ease of peacetime and wartime innovation, the special problems of technological innovation, and the role of intelligence in all of these. Some of these questions have been answered in individual chapters. Peacetime innovation has been possible when senior military officers with traditional credentials, reacting not to intelligence about the enemy but to a structural change in the security environment, have acted to create a new promotion pathway for junior officers practicing a new way of war. Wartime innovation, as opposed to reform, has been most effective when associated with a redefinition of the measures of strategic effectiveness employed by the military organization, and it has generally been limited by the difficulties connected with wartime learning and organizational change, especially with regard to time constraints. Technological innovation was not closely linked with either intelligence about the enemy, though such intelligence has been extremely useful when available, or with reliable projections of the cost and utility of alternative technologies. Rather, the problems of choosing new technologies seem to have been best handled when treated as a matter of managing uncertainty.

Can any other conclusions be reached by examining the three types of innovation together? Can any help be offered to Americans who are concerned with the possible need for military innovation in the 1990s? Taken as a whole, the study does provide a foundation for some conclusions about the role of resources, intelligence, and civilian control in military innovation.

Chapter 1 noted that studies of many different types of bureaucracies had found no relationship between levels of resources and the number of innovations an organization made. This book has not identified and studied every major innovation in the American military, but it is striking that the successful innovations examined were initiated in periods of constrained resources at least as often as in periods during which budgets were large and growing. The United States Navy transformed itself from a battleship navy to an aircraft carrier navy in a period in which naval budgets were modest and constrained by arms control agreements. The United States Marine Corps invented and developed a new form of amphibious warfare during the same period. The United States Air Force and the Joint Chiefs of Staff adopted a research and development strategy that provided the technology base for the Atlas ballistic missile in 1946, a time of rapidly shrinking budgets and demobilization. The few foreign cases studied also suggest that resource levels and innovation have no necessary connection. The components of the Royal Air Force responsible for the air defense of Great Britain showed remarkable foresight and competence in developing the aircraft and command network for their new mission during a period when Fighter Command was the poor sister, in terms of both resources and doctrine, to Bomber Command. Affluence was neither a barrier to nor a guarantee of innovation. The United States Army developed helicopter warfare capabilities during the 1950s, when it received unusually high levels of resources, but failed to create a genuine counterinsurgency capability during the same years.

The general lesson for students or advocates of innovation may well be that it is wrong to focus on budgets when trying to understand or promote innovation. Bringing innovations to fruition will often be expensive. Aircraft carriers, fleets of helicopters, and ICBM forces were not cheap. But *initiating* an innovation and bringing it to the point where it provides a strategically useful option has been accomplished when money was tight.

Rather than money, talented military personnel, time, and information have been the key resources for innovation. The study of peacetime military innovation showed that when military leaders could attract talented young officers with great potential for promotion to a new way of war, and then were able to protect and promote them, they were able to produce new, usable military capabilities. Failure to redirect human resources resulted in the abortion of several promising innovations. Human resources had to be treated with extreme ruthlessness in the United States Navy submarine forces in the Pacific in World War II; here innovation depended entirely on the character of the commanders of individual units.

Wartime innovation, even where successful, was less dramatic because of the lack of time for a more thoroughgoing implementation. Reformulating strategic measures of effectiveness was associated with allocating scarce resources to the tank in World War I and with assessing the impact of the United States submarine war against Japan, but the reformulation took a long time relative to the duration of the war. Peacetime innovation seems to have been relatively more important in large measure because of the time factor; time has not been available to assess wartime conditions, reformulate strategic conceptions, and build new forces before the outcome of the war is already largely determined. Under certain circumstances, associated with extremely decentralized control of operations (the United States Navy submarine force) or a decentralized capability to modify equipment in the field in response to enemy tactics (electronics warfare in the RAF in World War II), wartime innovation has played a larger role. Ordinarily, however, large military organizations are centrally directed, and reformulations of their concepts of operations and resource allocations must proceed from the top down. Strategic intelligence also tends to be analyzed at the top, even when control of operations is decentralized.

INTELLIGENCE AND INNOVATION

One of the more important findings is that for all three types, peacetime, wartime, and technological, intelligence about the behavior and capabilities of the enemy has been only loosely connected to American military innovation. It is by no means certain that this pattern has prevailed in other countries. There were factors that explain the relatively small role of intelligence in American peacetime and technological innovation that seem to be general, but not all were of this nature. Enemy plans and capabilities can be volatile and thus cannot provide the basis for innovations that takes years, if not decades, to bring to fruition. This instability would seem to affect every nation, although nations facing only one possible threat would have their intelligence task greatly simplified. Political barriers constrained intelligence collection and review for strategic target analysis in the Army Air Forces before World War II, but there is not reason to assume that the same problem would affect all countries. Constraints on the distribution of sensitive information restricted the role of intelligence in initiating and shaping innovations in the Navy during World War II and in the Air Forces after the war. Inadequate resources for data collection restricted the flow of technical intelligence about Germany to the United States Army before World War II. These difficulties might have been overcome by different policies, and thus similar problems may be more successfully handled by other countries.

Because of the theoretical and practical problems of utilizing intelligence about enemy plans and capabilities, peacetime and technological innovations in the American military have in practice been more closely linked to analyses of the anticipated security environment, which is determined by economic, technological, and political factors largely outside the control of either the United States or its potential adversaries. It was found to be more useful to focus on those determining aspects of the environment that were independent of the fluctuating policy decisions of the enemy. The emergence of the United States as a Pacific power, the potential of naval aviation, and the technological trends visible after World War II are examples of the way perceptions of change in the security environment provided the intellectual basis for successful, useful innovations.

Two methods for handling large, residual uncertainties have been employed in managing peacetime and technological innovations. Simulation of the impact of new capabilities has helped military leaders envision the shape the next war might take and how military innovations could affect it. Simulation was more important than intelligence collection and analysis because intelligence about the future is inherently problematic. Analysis of the current environment can narrow the range of possible futures, but imagining the future involves asking a series of "what ifs?" and thinking through the implications of the answers. Simulations have been useful in this process, but simulations are not predictions, and should not be treated as such. All simulation can do is identify a range of potential military requirements. In some cases, simulation and analysis was used to narrow the range of alternatives enough to permit the commitment of a service to the creation of a new capability.

Wartime innovation was hindered because of the difficulties in obtaining relevant, accurate information. In the cases of the tank and strategic bomber targeting, the lack of such data interfered with the development or successful implementation of the innovation. Spe-

cial arrangements for quickly deciding which new forms of intelligence are relevant to innovation in wartime, for obtaining and
analyzing that intelligence, and for rapidly transmitting it to those
portions of the organization responsible for new equipment and operations need to be anticipated if wartime innovation is to be more
successful. The normal arrangements have not worked well in support of wartime innovation.

CIVILIAN CONTROL AND INNOVATION

Another significant observation is that of the relatively minor
role civilian political leaders and scientists have had in the initiation
and management of military innovation. Political leaders in the
United States and Great Britain exercise effective control over many
military matters. They decide when to go to war, and determine
overall level of military budgets. In the cases studied in this book,
civilian political leaders have also had a great deal to say about how
resources should be allocated among the different sectors of the military, for example in the shift of resources from bombers to fighters
in the Royal Air Force or from fighters to bombers in the United
States Air Force.

Civilian political leaders, however, do not appear to have had a
major role in deciding which new military capabilities to develop,
either in peacetime or in war, although they did help protect or accelerate innovations already in progress in the cases of naval aviation and helicopter airmobility. The leaders of the United States and
Great Britain may have played a somewhat larger role in wartime,
creating new military capabilities that the military either had not
thought of or resisted. The role of David Lloyd George in the initiation of convoys in World War I is controversial, but might fall in this
category, as might the role of Winston Churchill in the development
of the tank. Civilians played a large role in the development of strategic targeting for the American bombardment of Europe and Asia.
The influx of civilians into the defense establishment after mobilization and the increased importance and visibility of technical issues
may contribute to this phenomenon. This study did find, however,
that many important wartime technical innovations, such as the
tank, the proximity fuse, and microwave radar, and organizational
innovations, such as new doctrines for submarine warfare and strategic targeting functions for American bombers, were pursued at the
initiative of military officers or with their vigorous support. War also

increased the political power of military leaders who developed mass popular support, and their increased power did hinder wartime innovation supported by civilians and other military leaders. Overall, the evaluation of the role of civilians in wartime innovation has to take account of the fact that peacetime innovations have been of relatively greater importance than wartime innovations. A larger role for civilians in wartime does not balance the larger role of military men in peacetime.

This judgment tends to cast doubt on the model of objective civilian control of the military advanced by Samuel P. Huntington.[1] Objective control operates by the formulation of clearly defined directives by civilians which are then left for the military to execute. If peacetime innovation operates through the mechanisms that determine the promotion of officers, civilian control would appear to require a hand in those mechanisms. In a professional military, this kind of civilian control is vigorously resisted, and Huntington is correct in noting that civilian intervention in the officer corps will tend to split it into factions that parallel those in civilian politics.[2] The French army before 1914 stands as a warning of how catastrophic such factionalism can become. A larger role for civilians in determining the rate and direction of peacetime innovation would seem to require finding a way to overcome the cost associated with an expanded role for civilians in the selection of officers for promotion. Direct confrontation with the military in this area is likely to result only in deadlock,[3] and some type of coalition between civilians and selected elements of the officer corps who are accepted as legitimate players in the promotion process would seem to be necessary. Precedents for such a coalition, however, are not evident in the cases studied in this book.

Independent civilian scientists have clearly been important in stimulating technological innovation by pushing back the frontiers of knowledge, most important in the case of nuclear fission weapons. They did not play a major policy role in pushing the American military to develop capabilities it otherwise would have neglected in the cases studied, and it is not clear that increasing their influence

[1]Samuel P. Huntington, *The Soldier and the State* (Cambridge: Harvard University Press, 1957), pp. 80–84.
[2]See, for example, John F. Lehman, Jr., *Command of the Seas: Building the 600 Ship Navy* (New York: Scribner's, 1988), pp. 37–38.
[3]Civilians confronting the military directly to overcome its professional self-interest and to promote innovation is advocated in connection with the development of ballistic missile defenses by Angelo Codevilla, *While Others Build: A Common Sense Approach to the Strategic Defense Initiative* (New York: Free Press, 1988), pp. 218–22.

in military research and development decisions would significantly change the pattern of technological innovation. The example of the independent role for American civilian scientists in military research and development in World War II suggests that they quickly become subject to the same organizational incentives that govern military officers and scientists in government.

INNOVATION AND THE FUTURE OF THE AMERICAN MILITARY

Many, if not most, of the analyses of the way in which the American military currently operates focus on improving the efficiency with which the services and the Department of Defense perform their existing functions.[4] For example, there have been steady searches in government, the academy, and the contract research world for less expensive ways of ensuring a survivable nuclear retaliatory force, defending West Europe against armored invasion, or maintaining control of sea lines of communication. Such studies are clearly in the interest of the United States, and the savings and increased security they could produce justify the effort. The task of identifying the need for new military functions and capabilities, however, is very different than the search for military efficiency. Thinking about peacetime military innovation requires a focus on the next twenty to thirty years. The United States military establishment focuses most intensely on the next annual budget, and the Department of Defense's Five-Year Defense Plan is regarded as long-term planning. This short-term focus will have to be overcome if the rate of peacetime innovation is to be increased.

The Soviet military, in the public writings of its senior officers, has focused on certain factors that resemble what this book has referred to as the technological dimensions of the security environment, and it appears to base the long-term development of its forces in part on these technology trends. Soviet military writers have identified three scientific-technical "revolutions," two of which are historical and

---

[4]There are countless studies suggesting ways to reform the American military to make it more efficient. A sample of recent publications includes Jacques Gansler, *Affording Defense* (Cambridge: MIT Press, 1989); Fen Hampson, *Unguided Missiles: How America Buys Its Weapons* (New York: Norton, 1989); Gary Hart and William S. Lind, *America Can Win* (Bethesda, Md.: Adler and Adler, 1986); William W. Kaufmann, *A Thoroughly Efficient Navy* (Washington, D.C.: The Brookings Institution, 1987); William Kaufmann and Jo Husband, *Defense Choices: Greater Security with Fewer Dollars* (Washington, D.C.: Committee for National Security, 1986).

one of which is occurring now. These revolutions occur not as the result of a single new weapon or technology, but when groups of technologies emerge that together transform the nature of warfare. Such a revolution took place in the 1920s and 1930s when the internal combustion engine, mobile radios, and military aviation combined to increase both the speed with which armies could advance and the depth to which they could penetrate. Another revolution took place in the 1950s, when ballistic missiles and nuclear weapons made it possible to bring overwhelming firepower down on the enemy even more quickly and to even greater depth. Beginning in the 1980s, Soviet military writers began referring to another revolution in military affairs, based on the development of military electronics, including computers, sensors, and communications systems. This revolution was producing a qualitative change in the effectiveness of nonnuclear weapons, making them as militarily effective as tactical nuclear weapons.[5] One cruise missile armed with nonnuclear munitions that are designed to find and kill military vehicles, Soviet writers have argued, can be as militarily effective as a one-kiloton neutron bomb. At the same time, these nonnuclear weapons carry with them far smaller military and political risks than nuclear weapons of comparable effectiveness. This revolution will take place over the coming decades and will transform warfare between advanced industrial states.[6]

If Soviet analysts are correct in their assessment, the security environment could undergo a structural change comparable to that associated with the development of naval aviation in the 1920s and 1930s or the nuclear revolution in firepower that led to the introduction of helicopter airmobility in the 1950s. The history of those innovations suggests that intensive simulation of alternative concepts of operation could help the American military choose the innovations that would take greatest advantage of the new military technologies. Once new missions and tasks were defined through these simulations, new career and promotion patterns would have to be established for the officers who specialized in the innovation. Activity within the military would shift to building these new capa-

[5]Notra Trulock III, Kerry Hines, Anne Herr, *Soviet Military Thought in Transition: Implications for the Long-Term Military Competition* (Arlington, Va.: Pacific-Sierra Research Corporation, May 1988), pp. 2, 28, 31.

[6]Ibid., p. 45; see also John Hines, Phillip Petersen, Notra Trulock III, "Soviet Military Theory from 1945 to 2000: Implications for NATO," *The Washington Quarterly* 9 (Fall 1986), 127; Phillip A. Petersen and Notra Trulock III, "A 'New' Soviet Military Doctrine: Origins and Implications," mimeo, Soviet Studies Research Centre, Royal Military Academy, Sandhurst, Summer 1988.

bilities. Since the resources available to the American military are not infinite, and indeed, are likely to steadily shrink in real terms in the period 1990–2000, this would mean fewer resources going to the maintenance of existing capabilities. This diversion of resources would, however, be small, since the emphasis would be on research, development, and simulated warfare for analytical purposes. Large uncertainties would exist about when these new capabilities would be needed for a war with the Soviet Union, or whether they would be needed at all. Research and development of a range of new military technologies, and mobilization plans for their mass production, rather than large-scale procurement of any one system, would appear to be the best way of managing those uncertainties, if the model of American guided missile development in the 1950s is applicable.

It is also necessary to look beyond war with the Soviet Union. The security environment is changing because economic growth in Asian countries has given them the industrial and technological basis for independent military power. India has developed nuclear explosives and rockets that could serve for intermediate-range ballistic missiles. Japan clearly has the economic means to become a great military power. Rates of economic growth in the People's Republic of China have been so high that even if they fall to half what they were in the 1980s the country will still have a gross national product as large as that of the Soviet Union during the decade 1990–2000.[7] Chemical weapons appear to be available to any country with a modest chemical industry. Large-scale shifts in the global distribution of power may be in progress that will force a change in the strategic perspective of the United States as large as the one associated with the nation's emergence as a Pacific power at the turn of the century. Huge uncertainties will be introduced if this happens. It is unclear which country, if any, among the new great powers we would have to oppose in war. It is unclear what military policy we would adopt if other powers went to war with each other.

A strategy that would prepare military innovations for this new world has to focus on the management of uncertainty, rather than on the construction of new capabilities tailored to predictions of what future wars would look like. It would be a mistake to construct a single scenario for a war with any of the new powers and then

[7]*The Future Security Environment* (Washington, D.C.: GPO, October 1988), report of the Future Security Environment Working Group, submitted to the Commission on Integrated Long-Term Strategy, pp. 4–5, 43–44, 47, 53.

build new capabilities based on that scenario. A more effective strategy would be to buy information about a range of uncertainties and then defer construction of new systems until new strategic requirements had become better defined. A necessary aspect of this program would be to make preparations for the rapid construction of the new systems.

This strategy for managing uncertainty has at least four concrete implications. First, we know little about the strategic perspectives of many of emerging powers. Compared with what we know about how the Soviet Union, Germany, or France thinks about war and strategy, we know little about the strategic cultures of Japan, India, or China. Strategic cultures are not like strategic plans. They are the result of political and cultural history and tend to be relatively stable over time. The study of these cultures would be inexpensive and could reduce our uncertainties about how these countries could use their new power.

Second, government money invested in Asian language training programs would be money well spent. Intelligence collection requires men and women with the necessary language skills. Without those capabilities, we will know less about strategic change in the new powers. When the Soviet Union emerged as a challenge to the United States in the 1950s, the United States government responded with a national-defense language program to teach students Russian. Analogous programs in Japanese, Hindi, and Chinese are necessary today.

Third, in conjunction with the strategy for managing the scientific-technical revolution, the United States needs to construct a wide range of scenarios and conduct imaginative conflict simulations in order to explore the shape of potential wars. We cannot allow ourselves to continue to confine military thinking about war to U.S.-Soviet nuclear exchanges and Soviet invasions of Europe. Wars between India and China, China and the Soviet Union, and a host of other possible combinations need to be explored to refine our understanding of the need for innovation.

Fourth, we must acknowledge that if war, whether with the Soviet Union or some new power, does increase in probability, we will need to be prepared to innovate rapidly to meet the new requirements. Historically, peacetime innovations have taken years to complete. Years may not be available when a new threat clearly emerges. A range of new capabilities needs to be thought through to provide alternatives from which policy makers can choose. This is difficult. What will be even more difficult is preparing the capability to rapidly build a new force once the choice has been made. Presently, knowledge of how we would rapidly mobilize the United States to create new forces is limited. In the 1950s and 1960s mobilization was thought of as a way of providing the material for a protracted world war with the Soviet Union. By 1964, very few Americans believed that this kind of war was likely, so mobilization studies went into decline. Today, the problem is different. We would be mobilizing to create new forces after a period in which they had been allowed to decline in size because there was no perceived need for a large standing force and no reliable way of knowing what the force should look like. How best to effect this kind of mobilization—from the production of equipment, to the creation of new military organizations, to the training of the officers who would command them—needs to be thought through.

If a war does occur, it is likely that some of our forces will be inappropriate to the realities of combat. The lesson of history is that wartime innovation is slow and difficult because of the need to redefine strategic measures of effectiveness and to collect and analyze the information relevant to creating new capabilities. Detailed study of the wartime process of collection, analysis, and use of intelligence might speed up the process of wartime innovation, but the historical precedents are poor.

Index

[263]

CORNELL STUDIES IN SECURITY AFFAIRS

edited by Robert J. Art, Robert Jervis, and Stephen M. Walt

Library of Congress Cataloging-in-Publication Data

Rosen, Stephen Peter, 1952–
  Winning the next war : innovation and the modern military /
Stephen Peter Rosen.
     p. cm.—(Cornell studies in security affairs)
     Includes index.
     ISBN 0-8014-2556-5 (alk. paper)
     1. United States—Armed Forces—History—20th century. 2. Great Britain—
Armed Forces—History—20th century. 3. Military art and science—United
States—Technological innovations—History—20th century. 4. Military art
and science—Great Britain—Technological innovations—History—20th
century. 5. Military art and science—United States—History-20th century.
6. Military art and science—Great Britain—History—20th century.   I. Title.
II. Series
UA23.R758   1991
355'.0332'0904—dc20   91-55235